KNOWLEDGE ACQUISITION IN CIVIL ENGINEERING

Prepared by the Committee on Expert Systems
of the Technical Council on Computer Practices
of the American Society of Civil Engineers

**Edited by Tomasz Arciszewski
and Lewis A. Rossman**

Published by the
American Society of Civil Engineers
345 East 47th Street
New York, New York 10017-2398

ABSTRACT

Expert systems are an intriguing approach for transferring valuable problem-solving skills from the minds of human experts to the logic of computer programs. For the past several years, intensive research and development activities into the fundamentals of expert systems and their application to civil engineering problems have been taking place. As more and more expert systems are built for practical applications, civil engineers need to improve their capability to acquire, assimilate, and codify knowledge that currently exists only in the form of personal engineering experience, judgement, or as site-specific information from individual projects. This book is written for all civil engineers who are interested in the role played by knowledge acquisition in expert system development. It is divided into three parts. Part I gives the reader a broad perspective of knowledge acquisition issues and an introduction to the basic techniques of both manual and automated knowledge acquisition. The second part consists of case studies concerning approaches to knowledge acquisition in three major civil engineering areas. The last part is an annotated bibliography of relevant literature in the field.

Library of Congress Cataloging-in-Publication Data

Knowledge acquisition in civil engineering / prepared by the
    Committee on Expert Systems of the Technical Council
    on Computer Practices of the American Society of Civil
    Engineers; edited by Tomasz Arciszewski and Lewis
    A. Rossman.
        p.   cm.
    Includes index.
    ISBN 0-87262-864-7
    1. Civil engineering—Data processing. 2. Knowledge
acquisition (Expert systems) I. Arciszewski, Tomasz. II.
Rossman, Lewis A. III. American Society of Civil Engineers.
Committee on Expert Systems.
TA345.K65   1992
624'.0285—dc20   91-41435
                                                           CIP

CONTENTS

PREFACE

"Experience is mortal, but learning knowledge will last forever."

Anonymous

Expert systems are an intriguing approach for transferring valuable problem solving skills from the minds of human experts to the logic of computer programs. The past several years have seen intensive research and development activities into the fundamentals of expert systems and their application to civil engineering problems. ASCE has organized and sponsored several conferences and symposia devoted to this new technology. It has also published two volumes of a monograph series entitled "Expert Systems in Civil Engineering." As a result of these activities, civil engineers have acquired a good understanding of the basic concepts and potential applications of expert systems within their profession. Many university researchers and consulting engineers are actively involved in building expert systems for practical applications. This has created a pressing need to improve our capability to acquire, assimilate, and codify knowledge that currently exists only in the form of personal engineering experience, judgement, and heuristics, or as site-specific information from individual projects. Knowledge acquisition has thus become a crucial area for insuring continued progress in the development and application of expert systems. Realizing the importance of this topic, the ASCE Expert Systems Committee decided in 1989 to add this book on knowledge acquisition to its monograph series.

The book is written for all civil engineers who are interested in the role played by knowledge acquisition in expert system development. The contributing authors include computer scientists, civil engineering faculty, and practicing engineers. This mix was chosen so that the reader could benefit from a more comprehensive view of knowledge acquisition in civil engineering, reflecting both recent theoretical developments in computer science as well as the practical experience gained by civil engineers building expert systems in the past few years.

The book is divided into three parts. Part I, Fundamentals, gives the reader a broad perspective of knowledge acquisition issues and an introduction to the basic techniques of both manual and automated knowledge acquisition. The second part, Case Studies, provides examples of knowledge acquisition approaches in three major civil engineering areas. Part III, Recommended Readings, is an annotated bibliography of pertinent literature that the interested reader can consult for further information and self-study.

Tomasz Arciszewski and Lewis A. Rossman, Editors

Acknowledgements

This book is an effort of the ASCE Expert Systems Committee to disseminate knowledge on expert systems to the civil engineeirng community. The book is third in a series, where each book deals with a specific issue related to expert systems. The first book, **Expert Systems for Civil Engineer: Technology and Application** was published by the ASCE in 1987, and the second book **Expert Syustems in Civil Engineering: Education**, was published in 1989. These books represent an ongoing effort of the Committee to prepare and carefully monitor a set of publications directly related to civil engineering and expert systems. Each chapter contributed to these books is reviewed by the Committee for scope and content. The editors appreciate the efforts of the entire Committee in guiding the development of this monograph and the technical support provided by Wayne State University in the preparation of camera ready manuscripts of four chapters.

Members of the Expert Systems Committee

Teresa M. Adams
Irtisham Ahmad
Robert Allen
Tomasz Arciszewski
David Ashley
Salah Benabdallah
William Bowlby
Stuart Chen
Louis F. Cohn
Michael Demetsky
Clive L. Dym
Gavin A. Finn
Bruno M. Franck
John Fricker
James H. Garrett, Jr.
Jesus M. De la Garza
John S. Gero
Geoffrey D. Gosling
Roswell A. Harris
Craig Howard
William Ibbs
Raymond Issa

Nabil A. Kartam
Simon Kim
Kincho H. Law
Raymond E. Levitt
Andrew B. Levy
Phillip J. Ludvigsen
Mary Lou Maher
Peter Mullarkey
Leonard Ortolano
Richard N. Palmer
William Rasdorf
Malcolm Ray
W.M. Kim Roddis
Paul N. Roschke
Timothy J. Ross
Lewis A. Rossman
Thomas V. Schields
Thomas Siller
Miroslaw J. Skibniewski
Duvvuru Sriram
Kenneth Strzepek
Iris Tommelein
Charles H. Trautmann
Terence A. Wiegel

Biographies of Authors

Tomasz Arciszewski is Associate Professor of Civil Engineering at Wayne State University. Previously, he taught at the University of Nigeria in Nsukka and at Warsaw Technical University in Poland. He earned his M.S. and Ph.D. degrees at Warsaw Technical University in 1970 and 1975 respectively. He is presently interested in the applications of various forms of inductive learning in civil engineering, particularly in structural optimization, decision making and the analysis of construction accidents. He gained practical design experience in Poland and Switzerland. He has authored or co-authored more than 50 publications in the areas of structural engineering, design methodology and artificial intelligence, including several book chapters. He is an Associate Editor of the ASCE Journal of Computing in Civil Engineering.

P.M. Berthouex, Professor of Civil and Environmental Engineering, the University of Wisconsin-Madison, has published a number of articles on wastewater treatment design and operation with a special interest in the application of statistical methods.

Harry Borchers is the executive manager of the North Penn Water Authority, Lansdale, Pennsylvania, with overall responsibility for all aspects of North Penn Water Authority activities.

Robert Clark is the director of the Drinking Water Research Division of the U.S. Environmental Protection Agency, in Cincinnati, Ohio.

Judith Coyle is an independent consultant on water and environmental subjects, in Ballwin, Missouri. She was formerly water quality manager and laboratory director for the North Penn Water Authority, Lansdale, Pennsylvania.

Steven Fenves is the Sun Company University Professor at Carnegie Mellon University, with appointments in the Department of Civil Engineering and the Engineering Design Research Center. He received his education in civil engineering at the University of Illinois (BS 1957, MS 1958, Ph.D. 1961). He served on the faculty at the University of Illinois until 1972, with visiting appointments at MIT, Cornell and the National University of Mexico. At Carnegie Mellon he has held the position of Head of the Civil Engineering Department and Director of the Design Research Center. His teaching, research and consulting have centered on the emerging discipline of computer-aided engineering. His work in this area began in 1962 with the development of STRESS, the first general-purpose structural analysis system. Subsequently, he has worked in network theory, representation of standards, engineering databases, knowledge-based expert systems, and Artificial Intelligence applications to design, synthesis and Machine learning. He is the author of 3 books and over 250 papers. His honors and awards include teaching awards from the University of Illinois (1962) and Carnegie Mellon (1989), the ASCE Walter L. Huber Research Price (1964), the University of Illinois College of Engineering Alumni Achievement Award (1984), election to the National Academy of Engineers (1976), election to Honorary Member of ASCE (1986), and the ASCE Moisseiff Award (1990).

Richard Forsyth is an ex-poet turned psychologist who now struggles to make a living as a computer expert. He holds a B.A. in psychology from Sheffield University and an M.Sc. in Computer Science from the City University in London. In 1984 he left his position as senior lecturer at the Polytechnic of North London to become a free-lance author and researcher specializing in machine intelligence and its applications. He has written and edited several books. Since 1989 he has been working as a senior research associate in the Psychology department at Nottingham University.

Jesus M. De La Garza is an Assistant Professor of Civil Engineering at Virginia Tech. His field of research is computer-aided construction with a current focus on computer-integrated design and construction through artificial intelligence software. He is also conducting research in the technology transfer field aimed at expediting the transfer of expert systems to the architecture- engineering-construction industry. Funding sources include the U.S. Army Construction Engineering Research Laboratory and the National Science Foundation. He teaches courses in construction engineering and management and a course on expert systems to graduate students in engineering.

Carlo Gavarini graduated in Civil Engineering at the University of Rome, Italy in 1958. He became full Professor in Structural Dynamics in 1975. His main research interests, for the past 15 years, lie in the field of earthquake engineering, an area in which he is involved both as a research and as a member of many National Committees.

Walter Grayman is principal of W.M. Grayman, Consulting Engineering, Cincinnati, Ohio specializing in computer applications in water resources.

Beth Hertz is water quality manager for the North Penn Water Authority, Lansdale, Pennsylvania, dealing with overall strategies for water quality assurance at NPWA, environmental issues, and customer education.

C. William Ibbs is a member of University of California's Berkeley Construction Engineering and Management Program, and has research efforts in the areas of project controls. Currently he is supervising Ph.D. students working in the computer-aided schedule generation; intelligent cost/ schedule exception report; construction work package definition; conceptual cost estimating and control; and contract strategy definition. He received support from such varied sponsors as the National Science Foundation, the Construction Industry Institute, IBM, Texas Instruments, Burroughs Corporation, Bechtel, Shimizu and the U.S. Army. He is a 1985 recipient of the NSF Presidential Young Investigator's Award (PYIA), and the Halliburton Foundation's "Excellence in Engineering Education" Award. He has chaired the ASCE Project Planning and Controls

committee for 4 years to supplement this research activity.

Wenje Lai is a Member of the Technical Staff, AT&T Bell Labs in Naperville, Illinois, where he works on information management systems and software development for telephone switching. He has a Ph.D. in Industrial Engineering from the University of Wisconsin-Madison.

Richard Males is the owner of RMM Technical Services, Inc., Cincinnati, Ohio, an independent consulting firm specializing in advanced computer technologies applied to water resources issues.

Kenneth Modesitt is the Department Head and Professor of Computer Science at Western Kentucky University. He had previously been a Professor at California State University at Northridge. While in California, he was employed concurrently at the Rocketdyne division of Rockwell International. In this capacity, he was involved in a multi-year effort to develop an expert system for test analysis of Space Shuttle main engine. Previously, he was employed by Texas Instruments, Inc. He has an extensive background in computer science and artificial intelligence dating back to his graduate student days at Carnegie Tech with Simon and Newell in 1965. He has published over 30 publications, and has presented tutorials throughout the United States and in Japan, Scotland, France, Sweden, Australia, and People's Republic of China.

Mohamad Mustafa is an engineer at Wayne County Public Department Engineering Design Division. Also, he is a Ph.D. candidate in the Civil Engineering Department, Wayne State University. He got his B.S. and M.S. degrees from the same institution. In 1985, he began his research on the methodology of inductive learning in structural engineering. This research produced results published in a book chapter, and in a journal and several conference proceedings papers.

Leonard Ortolano teaches courses in environmental policy implementation and water resources development in the Department of Civil Engineering at Stanford University. His current research focuses on environmental policy implementation in developing countries. He is also participating in a study to develop a "critiquing expert system" to aid operators of a water supply network serving the San Francisco Bay area. He is the author of Environmental Planning and Decision Making which was published by John Wiley and Sons in 1984.

Tommaso Pagnoni graduated in Civil Engineering at the University of Rome La Sapienza, in 1984. He obtained in 1988 an S.M. in Civil Engineering at MIT, where he is currently employed.

Catherine Perman currently works as a partner for Albathion Software in San Francisco, California. Albathion Software provides consulting services and software for expert systems and multi-media applications. She received her doctorate from Stanford University in Civil Engineering. Her research focused on developing and validating an expert system that diagnoses operating problems at wastewater treatment plants. Prior to studying at Stanford, She worked as an environmental engineering and systems analyst in the petroleum industry.

Yoram Reich is a recent Ph.D. graduate from the Civil Engineering Department at Carnegie Mellon University. He received his education in mechanical engineering at Tel-Aviv University (B.Sc., Summa-Cum-Laude, 1980; M.Sc. Magna-Cum-Laude, 1984). He practiced ship and steel structure design, and computer-design tool development for 6 years in Israel. His research is in the area of machine learning and knowledge acquisition for engineering applications, design methodologies, and advanced representations for finite-element and structural analysis. He has published 20 papers in these areas.

Lewis A. Rossman is an Operations Research Analyst with the U.S. Environmental Protection Agency's Risk Reduction Engineering Laboratory. He received a doctorate in Environmental Engineering from the University of Illinois at Urbana-Champaign, taught in the Civil Engineering Department at Worceser Polytechnic Institute, and has been with the U.S. EPA for the past 13 years. During this time he has developed numerous decision support models for a variety of problems in water quality and hazardous waste management. He has also contributed to EPA's expert systems for evaluating synthetic membrane liners and closure plans for hazardous waste management facilities.

Zahra Tazir received her degree in Civil Engineering at the *Ecole Nationale Polytechnique d'Alger* in 1984. She attained an S.M. in Civil Engineering at MIT, where she is now completing a Ph.D. program. Her research interests lie in the fields of earthquake engineering and knowledge based applications.

Wojciech Ziarko is an Associate Professor of Computer Science at the University of Regina, Saskatchewan, Canada. He studied at the Faculty of Fundamental Problems of Technology at Warsaw Technical University, where he received his M.Sc. degree in applied mathematics in 1975. He continued his studies at the Institute of Computer Science of the Polish Academy of Sciences, where he received his Ph.D. in Computer Science in 1980. His earlier research interests included theory of databases and information storage and retrieval. His present research is concentrated on artificial intelligence, particularly the area of adaptive systems and the theory and knowledge-discovery application of the mathematical model of rough sets. He has authored and co-authored over 40 research papers.

CHAPTER 1

In Search of Knowledge

Richard Forsyth

1. Knowledge, Intelligence and Learning

Since the early days of our civilization the human race has been engaged in a search for knowledge. Knowledge has meant the chance to survive and prosper. It has been a subject of keen interest to philosophers, scientists and engineers. The importance of knowledge in our times, often called the age of the "knowledge revolution" cannot be exaggerated. Knowledge acquisition and its use in expert and knowledge-based systems, is becoming an important part of engineering.

In this chapter, we will concentrate on an engineering approach to knowledge. We will assume that knowledge in a given domain is a consistent system of interrelated facts and processes relevant to that domain. This knowledge can be divided into procedural and conceptual knowledge. Procedural knowledge is usually tacit: it consists of ways and means of attacking problems learned by practitioners largely through their own experience. Conceptual knowledge, on the other hand, typically exists in symbolic or propositional form, e.g. in textbooks, as well as in the heads of the experts.

Humans seem to be capable of working effectively with fragmentary and sometimes inconsistent knowledge which machines cannot handle. Expert knowledge is hard to represent in a form that can be used by computers. Therefore a process of knowledge acquisition and refinement is necessary to transform an individual's knowledge of facts and processes into a usable format.

We are interested in knowledge in order to utilize it in decision support systems, which will be used over periods of time. Our world is rapidly changing, and so are demands on our systems. Ideally, they should be able to adapt to changing conditions and modify their knowledge bases. This ability is often referred to as machine intelligence. But what is really understood here by machine intelligence? How can this type of intelligence be compared with human intelligence? What indeed is intelligence? We will attempt to answer these crucial questions first, at least within the context of engineering. Our answers should also explain why machine learning is of growing importance and why there is an evolution of knowledge acquisition from human to machine to integrated learning.

For our purposes we will assume Hofstadter's (1985) definition of intelligence, which can be used with humans and other systems. This defines intelligence by five abilities:

1. to respond to situations flexibly;
2. to make sense of ambiguous or contradictory messages;
3. to recognize the relative importance of different element of a situation;
4. to find similarities between situations despite differences that may separate them;
5. to draw distinctions between different situations despite similarities that may link them.

These abilities taken together represent an ability to adapt to changing conditions, or simply an ability to learn. Human and machine intelligence are both understood by us to be based on learning ability. Learning produces new knowledge and modifies existing

1

knowledge. Human learning is the key to human intelligence. By analogy, we consider machine learning to be key to machine intelligence.

Learning has been the subject of investigation by psychologists for over a century. There are two main approaches to the psychology of learning: behavioral and cognitive.

Behaviorists regard the organism as a black box and consider only its inputs (stimuli), outputs (responses) and the relationship between them. Most of their learning experiments have been conducted with rats and pigeons. The objective of such experiments is frequently a determination of the stimulus-response relationship for repeated simple trials. This relationship, sometimes expressed graphically in a <u>learning curve</u>, relates the probability of a correct response by the tested organism to the number of tests conducted. Interestingly, learning curves of a similar shape to those found in animals have been reported for experimental machine learning systems. For example, Fig. 1 compares a typical learning curve for a living organism (Fig. 1.a) with an example of a relationship (Fig. 1.b) between number of examples and accuracy of the case-based predictions of optimal cross-sections in a rigid steel frame, obtained by Arciszewski and Ziarko (1991).

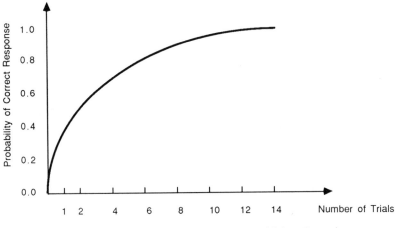

Figure 1. (a) Learning Curve for a Living Organism

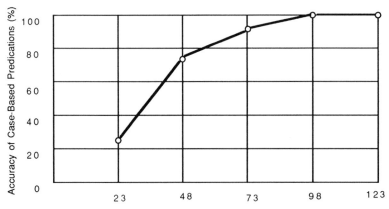

Figure 1.(b) Relationship between Accuracy and Number of Examples for Case-Based Prediciitons of Optimal Cross Sections in a Rigid Steel Frame (Arciszewski and Ziarko, Structural Optimization: Case-Based Approach, Journal of Computing in Civil Engineering, Vol. 5, No. 2, April, 1991. Reprinted by Permission of the American Society of Civil Engineers)

Cognitive theorists are concerned with conceptual structures constructed by the nervous system, as well as with overt stimuli and responses. They typically conduct their experiments with people, but usually on quite constrained tasks, which deal with only part of the phenomenon of learning as defined above.

At present there is no unified theory of learning and memory in psychology, combining both behavioral and cognitive approaches. So psychologists can provide only tentative clues and ideas on how to devise machine learning systems. For that reason, most of the advances in the field of machine learning to date have come from research in the field of Artificial Intelligence, or AI. In fact, psychologists take their theories of human learning from computational models more often that AI workers base their simulation on psychological theories.

2. Historical Review

To understand the history of AI, two basic types of computer architecture have to be identified. Von Neuman architecture is still dominant. In this case, all calculations are sequentially conducted by a single processor using a separate memory. The Connection Machine has a different architecture. It is a parallel computer with a large number of simple processors, each with its own memory. All calculations are concurrently conducted by individual processors working on various parts of the same problem. The architecture of the Connection Machine can be simulated on a Von Neuman computer, and one such implementation of the concept of concurrent computing is called "neural network."[*]

It is possible to identify five boom-then-bust episodes in the history of AI which, for the sake of simplicity, it can be divided into five decade-long chunks in Table 1.

Table 1. Short History of Artificial Intelligence

Decade	Label	Main Concern
1950s	The Dark Ages	[Neural Networks]
1960s	The Age of Reason	[Automated Logic]
1970s	The Romantic Movement	[Knowledge Engineering]
1980s	The Enlightenment	[Machine Learning]
1990s	The Gothic Revival	[Neural Nets Revisited]

This summarizes the "main stream" of AI development, though, of course, it should be remembered that AI researchers do not change the whole thrust of their work on 31st December of every tenth year. It should also be remembered that there are in every phase a handful of mavericks working outside the main conceptual framework -- either using a previously discarded approach or one whose time has yet to come (which sometimes amounts to the same thing).

As we enter the 1990s a counter-reformation is underway within AI. Research students and grants are being diverted from the symbolic knowledge-based paradigm which held sway in the 1970s and 1980s towards neuro-computing connectionism. I call this a "Gothic revival" because we have now come full circle: neural-net computing was the guiding theme of the 1950s (Selfridge 1959).

Extravagant claims are being made about the potential of this new (or newly rediscovered) technology, and the race is on, especially in the US and Japan, to become world leader in neuro-computing.

Why should the neuro-computing approach succeed now when its limitations were exposed more than 25 years ago? After all, though there has been progress in neurophysiology, there has been no tremendous breakthrough in our understanding of the nervous system during the intervening years.

The answer is twofold. In the first place, modern computing machinery is many orders

---

[*]The concept of neural networks is also discussed in the chapter "Machine Learning in Knowledge Acquisition."

of magnitude more powerful than anything available to the cyberneticians of the 1950s, and computer power is still increasing relentlessly. It is possible to construct neural nets with hundreds of thousands of processing elements which, while not rivalling the capabilities of higher mammals, do approach or even exceed the complexity of the brains of insects, molluscs and other "lowly" creatures.

Secondly, and perhaps more important, learning rules for multi-layered neural networks have been discovered. A particular weakness of the Perceptron (Rosenblatt 1958) -- identified by Minsky and Papert (1969) when they delivered the <u>coup de grace</u> to earlier work on neural networks -- was the elements. This was not because multi-layer systems have no advantages over single-layer systems (they have) but because the error-correcting rules of Rosenblatt (1962) and Widrow & Hoff (1960) could not be guaranteed to converge on a solution when used with multi-layered systems.

In the 1980s a number of workers independently discovered what has become known as the <u>back-propagation</u> rule, which overcomes this difficulty (Rumelhart & McClelland, 1986). Perhaps the best known success of back propagation is Sejnowski's NETtalk (Sejnowski and Rosenberg 1987), a reading system which learns to pronounce English text.

Thus the connectionist revival is essentially a reaffirmation of the importance of machine learning to machine intelligence.

3. Explicit versus Implicit Knowledge

Neural computing systems are extremely interesting, and it is true that they were unjustly neglected for more than twenty years, but they are not the answer to every computing problem. It will be a great waste of energy if AI becomes a battleground between the connectionists and the symbolists; although there are signs that this is happening.

The great advantage of connectionist systems is their adaptability: a neuro-computing system can, in theory, be trained to perform any desired input-output mapping. The advantage of symbolic knowledge-based systems, on the other hand, is that they make expertise explicit. The battle (if it is a battle) is over the importance of explicitness, since, in general, the knowledge acquired by a neuro-computing system is totally inscrutable. The question which really divides the two camps is: can we trust an expert system whose expertise cannot be understood by people?

Imagine an air traffic control system based on a simulated neural network that has been taught to recognize and predict impeding mid-air collisions. We might want to use it if we had good evidence (e.g. from a parallel trial) that it was more reliable than human operators; but we would not be very happy to stop at that. Undoubtedly we would seek ways of "de-compiling" its knowledge into some format that humans could inspect. In short we would carry on the process of turning implicit know-how into explicit knowledge.

Thus we must recognize that <u>both</u> trainability <u>and</u> the explicit representation of knowledge are desirable goals.

4. Connectionist Models of Memory

Human memory is a remarkable phenomenon, much studied but little understood. From an engineering viewpoint, we simply do not know enough to build a system that replicates its functionality; but connectionist researchers have made some progress in that direction.

<u>WISARD</u>

One interesting memory model is the WISARD visual recognition system (Aleksander & Burnett 1984). This is a parallel computer with no processors, only RAM (random access memory).

WISARD works by scanning a TV image that has been digitized into 512 x 512 pixels. Each pixel can be only on or off. The pixels are sampled in groups: eight pixels are randomly (but repeatably) assigned to each feature detector. A given pixel can be concurrently monitored by several feature detectors. When a group of 8 pixels is considered, the feature detector monitoring the group must be in one of 256 possible states, since the power of a Cartesian product of 8 binary sets is $2^8 = 256$. There are 32768 feature detectors working in parallel. Each detector is allocated its own RAM - bank of 256 locations, so it can store all possible combinations of states of the pixels monitored. Prior to training, all bits in all the RAM-banks are set to zero. During training, each detector monitors individual pixels in its group and stores a 1 at the specified address of its RAM-bank if the image is an instance of the class it is being taught to recognize.

Afterwards, when a new image is presented, each detector retrieves a bit from its RAM-bank according to the address pattern on its input lines. These are added and the total is used as a confidence measure that the image is of the target class. A high total shows that many detectors are in the same state as when a positive training instance was displayed during training.

WISARD, in effect, computes a fingerprint of what it is looking at and compares it with past experience. Its has been used commercially for factory inspection tasks, where high-speed pattern recognition is particularly valuable. It is a system that attacks complexity with quantity, rather like the human brain, and has proved surprisingly resistant to distortion of the image.

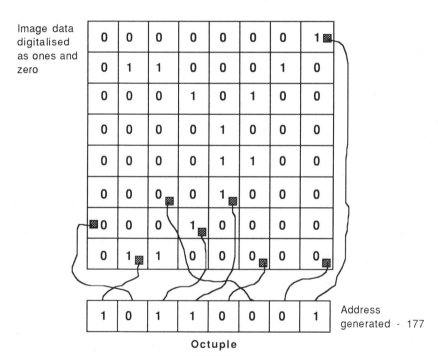

Figure 2. WISARD Schematic

Perhaps the most significant thing about WISARD is that it does not need to be programmed. If we want it to distinguish a nut from a bolt on a moving conveyor belt, there is no need to analyze the shapes of nuts or bolts, we just let the machine work out the difference for itself. Then if we want it to distinguish spanners from screwdrivers, we just re-train it.

CMAC

Another important distributed memory system is CMAC (the Cerebellar Model Articulated Controller), devised by James Albus (1971; 1981). This system was inspired by studies of the cerebellum, that part of the brain which deals with balance and motor coordination. It has been applied to several tasks, mostly in the area of robotic control.

CMAC can be viewed as a gigantic look-up table. A vector of sensory input values defines the current situation. This vector is used for addressing a table in which the appropriate response of the robot to that situation is stored. Of course a huge table in which every possible input pattern mapped to a unique location would be useless. Firstly, it would be practicable. Secondly, it would not generalize: the robot would need to be trained on each distinguishable situation.

CMAC overcomes the storage-size problem, and at the same time achieves automatic generalization, by mapping from an input vector to a set of addresses rather than to a single address. The mapping is arranged so that similar input vectors give rise to sets with overlapping membership. The closer the two input situations are, the more addresses they have in common, and so the more similar their output is likely to be.

Once a set of addresses has been computed from the quantized input values, the output is determined simply by summing the weights held at those addresses in the main memory table. Learning is accomplished by altering those weights. If the response (R) is different from the desired value (P) by more than a small tolerance then Delta is added to every weight that was summed to produce the response. Delta is calculated from the formula

Delta = G * (P - R) / A

where A is the number of addresses in the address set and G is a learning-rate parameter, such that $0 < G < 1$.

Rote memory may not seem very exiting at first glance, but CMAC's power of generalization in a noisy environment is something that has eluded more sophisticated AI systems, and Albus has shown that CMAC can be taught input-output mappings of arbitrary complexity to any desired accuracy level. Moreover, unlike some other neural-net models, it can operate in real-time on standard hardware.

5. Learning and Knowledge Acquisition

CMAC is an effective learning device, which will undoubtedly have applications in several branches of engineering, but its "knowledge" is quite opaque. During training its performance improves markedly, but to an outside observer all that has happened is that a number of weights in a large look-up table have been changed. Like some human experts it is inarticulate: it can perform well but cannot explain its actions. Much the same applies to WISARD, and to most connectionist learning systems.

For some purposes this is adequate. We just treat the machine as a black box and do not inquire into its inner workings. But for an expert system this is not good enough, because an expert system must be able to explain its own reasoning.

This is one reason why neuro-computing will not eliminate the need for symbolic machine learning, where the whole object of the exercise is to arrive at compact symbolic descriptions -- e.g. decision rules. The value of symbolic rules such as

    IF mean_corpuscular_volume is high
    AND gamma_globulin_level is high
    THEN liver_damage is likely

    or

    IF minimum_temperature_celsius > rainfall_mm
    AND maximum_temperature_celsius > 11.4
    AND sunshine_hours > 8.27
    THEN rain_probability_tomorrow is 0.07

is that we can understand them, we can criticize them and we can teach them to other people.

Systems that create symbolic decision rules of this sort have been available for a number of years. (See Forsyth & Rada 1986; Forsyth 1989; Lenat, 1983; Michalski & Chilausky 1980; Quinlan, 1986; Wong & Ziarko 1986.) A number of such systems have been applied to serious problems, as set out in Table 2, which is far from being a exhaustive list.

Table 2. Applications of Symbolic Learning

Problem	Reference
Glass Identification	Evett & Spiehler (1988)
Morphine Overdose Response	Spiehler (1988)
Polyethylene Production	Havener & Tischendorf (1989)
2D Image Recognition	Shepherd (1983)
Soybean Disease Diagnosis	Michalski & Chilausky (1980)
Jet Engine Test Analysis	Asgari & Modesitt (1986)

6. The Human Use of Machine Learning

Machine learning, therefore, is a viable technique in engineering and other disciplines; but it must be used with care.

Most inductive systems work in the following way:

Given

a collection of instances
described on a number of attributes
with correct class membership known

Find

a rule expressed in terms of those attributes
that assigns as many instances as possible
to the correct category.

Thus nearly all symbolic machine learning algorithms generate classification rules from examples. This means that they share certain common features, and makes it possible to lay down guidelines for their use in knowledge acquisition. (See, for example, Hart 1986; Kidd 1987.)

General Guidelines

Machine learning packages can be usefully applied to a wide variety of tasks, provided they are used in a methodical manner, as outlined below.

1. Define the goal of the project.
2. Select a suitable algorithm or package.
3. Collect a representative sample of training cases.
4. Present the data in a format suited to the induction system (which involve editing or re-coding).
5. Run the induction program.
6. (a) if results are very poor, quit.
   (b) if results are promising, revise the attributes and/or collect more training data and iterate from step 4.
   (c) if results are satisfactory, continue with step 7.
7. Check the rules on fresh unseen cases, and attempt to estimate the likely error rate using statistical methods. (Validation, phase 1.)
8. Subject the rules to human scrutiny for acceptability in the real world. (Validation, phase 2.)
9. Start using the rules in the field, beginning with a parallel trial. (Validation, phase 3.)

None of these stages is necessarily easy, but perhaps the one that causes most problems is step 6. The correct action at step 6 depends on the answers to at least three difficult questions:

> How good is good enough?
> When is more data needed?
> Do we need more attributes?

How Good is Good?

It is surprisingly easy to embark on a machine learning project with only a vague idea of how to measure success. Typically there will be a conventional system (human or computerized) for the same task which establishes a reference level of performance; but that performance level may turn out on closer inspection to be multi-faceted.

For example, suppose the goal is to predict the next day's weather conditions at London's Heathrow Airport. It may seem obvious that the standard to beat is the forecast of the Meteorological Office. If the rules generated by a learning system give correct predictions 80 times out 100 days while the public forecasters (aided by their supercomputers) are correct 75 times in 100 days then the rules surely look superior. But is a difference of 5% significant? Is 100 days a long enough testing period?

In fact, there is a standard statistical formula (see, for instance, Bruning & Kintz 1968) for deciding such questions:

$$z = (P - p) / sqrt((P*(1-P) + p*(1-p)) / (N + n))$$

Here P and p are two proportions and N and n are the number of cases in the two groups. The result (z) is a standard score which can be referred to the normal distribution to assess the significance of the difference between the two proportions. Using the (imaginary) weather-forecasting frequencies quoted above, the z score works out at 1.1995, which is a result that could be expected by chance alone with a probability of about 0.38. Thus it would not be regarded by statisticians as significant. Unfortunately, even such elementary statistical testing is rarely used to evaluate the results of knowledge acquisition projects.

Other common difficulties connected with comparative performance evaluation are:

(a) the rules give more false positives but the conventional system gives more false negatives (or vice versa);
(b) the rules are better under some conditions, but the conventional system is better in other (e.g.) better in winter than summer, better at predicting fog but worse at predicting windspeed);
(c) the rules are cheaper to apply, but not so accurate;
(d) the rules are more often right but make bigger errors when they are wrong.

The learning system itself cannot resolve such dilemmas. In fact, most learning packages make it quite hard even to address trade-offs such as these. But unless the user can resolve them (by some kind of cost-benefit analysis), the learning system is practically useless.

Adding Data

The question of when to gather more data is another that cannot be given a general answer. It has been considered in the statistical literature (e.g. James 1985) but is often ignored by AI workers. The fact that the standard error of an estimate is inversely proportional to the square root of the number of cases involved gives a rational basis for deciding how much data to gather. Siegel (1956), in his textbook of non-parametric statistics, argues from this fact that researchers should aim to collect sufficient cases so that

$$N * P * (1-P) > 9$$

when N is the total number of cases and P is the proportion of events counted as success.

Thus if an event occurs with a probability of 0.16 this formula requires that

$$N > 9 / (0.16 * 0.48) > 66.96$$

so at least 67 cases need to be investigated.

However, the value of this rule has not been widely accepted; and if we insist on reducing the probability of false positives or false negatives (whichever is more costly) below a set threshold we may well find ourselves forced to gather a huge amount of data -- which is expensive and time-consuming.

Probably the most valid general rule of thumb is: collect at least twice as much data as you first think you will need.

Adding Attributes

Adding a new column to a data file is usually a tedious chore (which may well involve re-investigating cases that have already been measured). As far as possible it should be avoided. This is where advice from a real human expert is crucial. With expert advice at the problem-formulation stage, this problem should never arise, because you should have a good set of discriminatory variables from the outset.

However, sometimes even with good advice, you can end up with a large and bushy decision tree or an over-complex yet ineffective rule. This may well indicate that important attributes are missing, though it may also indicate that the whole problem is ill-conceived. The only way to tell which applies in your case is to go back to the domain expert and re-examine both the rules and the data. If, in consultation, you decide that the addition of some more attributes will solve the problem, it is worth trying; if not, it is probably time to quit and move on to something else. Sometimes cutting your losses is the correct decision. Machine learning cannot solve every problem.

But whether the rules are good or bad, there is one piece of advice too easily forgotten:

DO NOT STOP WITH THE RULES.

They are just the starting-point for further investigation. You will not get full value from them unless you ask questions like

why does it use variables A and B but not X?
what types of cases does it still get wrong?
could the rules be expressed in a shorter form?
how different are they from what we expected?

and go on to answer them. In a successful machine learning project the user should always learn more than the computer.

Appendix - References

Albus, J., (1971). "A Theory of Cerebellar Function," *Mathematical Biosciences*, No. 10, pp. 25-61.

Albus, J., (1981) *Brains, Behavior & Robotics*, Byte Books, Peterborough, NH.

Aleksander, I., Burnett, P., (1984). *Reinventing Man*, Pelican Books, Middx.

Arciszewski, T., Ziarko, W., (1991). "Structural Optimization: A Case-Based Approach", *Journal of Computing in Civil Engineering*, Vol. 5, No. 2.

Asgari, D., Modesitt, K., (1986). "Space Shuttle Main Engine Test Analysis, a Case Study for Inductive Knowledge-Based Systems Involving very Large Databases". *Proc, IEEE*, pp. 65-71.

Bruning, T., Kintz, B.L., (1986). *A Computational Handbook of Statistics*, Scott, Foresman & Co., Glenview, Illinois.

Evett, I., Spiehler, E., (1988). "Rule Induction in Forensic Science," *Knowledge-Based Systems in Administration & Government* (Ed. Pafter) CCTA, London.

Forsyth, R., (ed.), (1989). *Machine Learning: Principles and Techniques*, Chapman & Hall,

London.

Forsyth, R., Rada, R., (1986). *Machine Learning*, Ellis Horwood, Chichester.

Hart, A. (1986). *Knowledge Acquisition for Expert Systems*, Knogan Page, London.

Havener, J., Tischendorf, M., (1989) "Use of Genetic Algorithms with PC/BEAGLE," *Eastman/Kodak AI Newsletter*, Summer-89, pp. 5-6.

Hofstadter, D., (1985). *Metamagical Themas*, Viking Press, NY & London.

Kidd, A., (ed.), (1987). *Knowledge Acquistion of Expert Systems*, Plenum Press, New York.

Lenat, D., (1983). "Eurisko: a Program that learns new Heuristics and Domain Concepts," *Artificial Intelligence*, No. 21, pp. 61-98.

Michalski, R., Chilausky, R.L., (1980). "Knowledge Acquisition by Encoding Expert Rules versus Computer Induction from Examples, a Case Study Involving Soybean Pathology, *Int. J. Man-Machine Studies*, No. 12, pp. 63-87.

Minsky, M., Papert, S., (1969). *Perceptrons: An Introdiction to Computational Geometry*: MIT Press, Boston.

Quinlan, J., R., (1986). "*Induction of Decision Trees*" Machine Learning, No. 1, pp. 81-106.

Rosenblatt, F. (1958). "The Perceptron, A Probabilistic Model for Information Storage & Organization in the Brain", *Psychological Review*, No. 65, pp. 386-404.

Rosenblatt, F. (1962). *Priniciples of Neurodynamics*, Spartan Books, New York.

Rumelhart, D. Mcclelland, J. (eds.), (1986). *Parallel Distributed Processing*, vols, 1 & 2: MIT Press, Cambridge, Mass.

Sejnowski, T. & Rosenberg, C.R., (1987). "Parallel Networks that Learn to Pronounce English Text," *Complex Systems*, No. 1, pp. 145-168.

Selfridge, O. (1959). "Pandemonium: a Paradigm for Learning," *Mechanization of Thought Processes*, HMSO, London.

Shepherd, B., (1983). "An Appraisal of a Decision-Tree Approach to Image Classification," *Proc. 8th IJCAI*, Karlsruhe, pp. 473-475.

Siegel, S., (1956). *Nonparametric Statistics for the Behavioral Sciences*, McGraw-Hill, NY.

Spiehler, V. (1988). "Computer Assisted Interpretation in Forensic Toxicology", *TIAFT Proceedings*, California.

Widrow, B., Hoff, M., (1960). "Adaptive Switching Circuits," *Inst. Radio Engineers Western Electronics Convention Record*, part 4, pp. 96-104.

Wong, S.D.M., Ziarko, (1986). "Algorithm for Inductive Learning", *Proc. Polish Academy of Sciences*, No. 34, pp. 271-276.

CHAPTER 2

Basic Principles and Techniques in Knowledge Acquisition

Kenneth L. Modesitt

A goal of life is to acquire <u>wisdom</u>, not just knowledge...
Wisdom is the right use of knowledge.
To know is not to be wise.....
To know how to use knowledge is to have wisdom

Spurgeon
New Dictionary of Thoughts, 1965

1. Introduction

Where does knowledge acquisition fit in the scheme of expert system development (or knowledge base systems -- KBSs, as the author prefers to call them)? A simple graphic gives one perspective. See Fig. 1. As can be seen, a KBS cannot exist until expert knowledge from <u>some</u> source is acquired or elicited. It is the highlighted box which is the primary subject of this monograph, in the domain of civil engineering.

The purpose of this chapter is to discuss knowledge acquisition in some depth. We will try to articulate some basic principles and techniques of knowledge acquisition which have worked well in the past. In particular, we will explore those which have involved human beings as the intermediary. This intermediary, usually called a "knowledge engineer," is the agent for the transmission of some type of knowledge from a knowledge source to a destination, normally with considerable "noise" added during the transmission. The next chapter will also discuss knowledge acquisition, but from the view point of a computer program being the Intermediary. That is, automated knowledge engineering tools will be the focus there, whereas here we are more concerned with manual tools for knowledge engineers. See Fig. 2 for some sample knowledge sources and such intermediaries (McGraw and Harbison-Briggs 1989).

This chapter presents a holistic perspective on knowledge acquisition, rather than one listing a set of "recipes." The field of knowledge acquisition is a <u>very</u> long way from providing a simple list of directions! Consequently, considerable discussion is given to the "why" and "what" of knowledge acquisition before delving into the "how."

<u>Why</u> are we interested in acquiring knowledge? I believe it is because we wish to do a better job of solving problems -- of filling gaps between what is currently and what could be in the future by changing our heuristics over time. We wish our functional needs to be satisfied effectively (with high quality) and efficiently (with minimum expense).

The "what" of knowledge acquisition is first investigated from the context of various types of knowledge which can be acquired. We discuss six categories: knowledge, comprehension, application, analysis, synthesis, and evaluation. Then an earlier definition for knowledge acquisition is extended: "the translation and transformation of problem solving expertise from a knowledge source (e.g., human expert, documents) to a human or computer program destination." Common examples are drawn from the fields of education and work (training and performance).

11

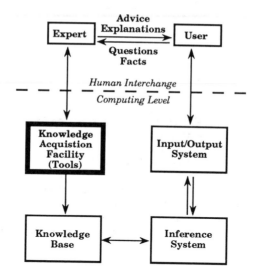

Figure 1. The Architecture of a Knowledge Based System.

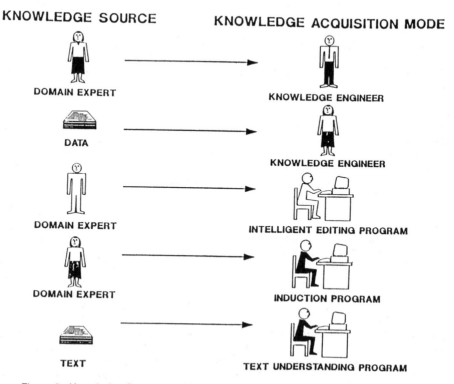

Figure 2. Knowledge Sources and Intermediaries.
(Karen L. McGraw/Karan Harbison-Briggs, KNOWLEDGE ACQUISITION:
Principles and Guidelines, (C) 1989, p. 10. Reprinted by permission of
Prentice Hall, Inc., Englewood Cliffs, New Jersey.)

The "hows" of knowledge acquisition are undertaken next, from both the viewpoints of humans and computers, as acquirers of knowledge. For each, the relevant methods are divided into past, present, and some future ones. Time itself is also seen to be critical over the lifetime of a learning entity, as progress is made from being a student or novice to becoming a knowledgeable and hopefully wise expert.

The "practical" part of this entire chapter is contained in a "heuristic cookbook," drawing upon the key work of *Knowledge Acquisition: Principles and Guidelines* (McGraw and Harbison-Briggs 1989). In this part, the major techniques which are currently used by knowledge engineers are summarized and compared. The importance of validating the resulting systems is also covered.

Future issues in knowledge acquisition include the role of <u>cooperative</u> systems (cooperative people, cooperative computers, and cooperative people <u>and</u> computers). Other issues address the difficulty (impossibility?) of a machine having all the senses which a human being uses in her learning, as well as an impetus to learn and freely ask questions.

The chapter concludes with a discussion of "so what?" What difference does all this material about knowledge acquisition make? Is the world any better now that it was before? It is the hope of this author that you will answer "yes" to the question. And with that optimism in mind, please join in the journey coming up!

2. Why Do We Wish to Acquire Knowledge?

To help solve major problems of today and tomorrow with technological solutions. To satisfy certain needs by filling the gaps between what currently is and what should be tomorrow.

In the context of this monograph, we address primarily the filling of gaps by technological solutions directed toward functional needs, such as shelter, transportation, energy, communication, etc. rather that the ones defined by Maslow. See Fig. 3.

Actually, we are far more concerned <u>not</u> with the acquisition of knowledge, but with the "application" of it. And it is not the application of just any old knowledge -- it is the application of the newest useful knowledge. This ties in with the definition of the engineering method as given by Dr. Billy Koen, a noted nuclear engineer. Koen says the engineering method is:

```
NEEDS ARE RELATIVELY INVARIANT IN TIME, PLACE, &
   CULTURE
SOLUTIONS USUALLY VARY RELATIVELY RAPIDLY
"GOOD" ONES ARE MULTIPLE-PURPOSE, OFTEN SIMPLE, SMALL,
SAFE, INEXPENSIVE, AND GENERALIZABLE

EXAMPLE NEEDS          EXAMPLE SOLUTIONS

DEFENSE                SPEAR, MACE
FOOD                   FARMING, WELL, FIRE
SLEEP                  NO-DOZE, WATERBED
SHELTER                QUONSET HUT, MANSION
CLOTHING               LOINCLOTH, 3-PIECE SUIT
ENERGY                 WINDMILL, SOLAR PANEL,
TRAVEL                 BICYCLE, SPACE SHUTTLE
HEALTH                 ASPIRIN, CAT SCAN
ENTERTAINMENT          YOYO, MONTE CARLO
COMMUNICATION          WRITING, DIRECT BROADCAST SATELLITE
PROCREATION            PILL (BOTH KINDS)
CURIOSITY              "NONE", EXCEPT LIFE AND THE WORLD
LOVE OF SELF
& OTHERS               ??
CALCULATING            ABACUS, COMPUTERS
```

Figure 3. Sample of Functional Needs and Technological Solutions.

"the strategy for causing the best change in an uncertain or incomplete situation within available resources" (Koen 1985).

Koen goes on to say that the use of heuristics is key to the engineering method. The well-known mathematician, George Polya, has made substantial contributions to the field of heuristics in the mathematical domain. Even though somewhat dated, the work on *How to Solve It* remains a classic today (Polya 1973). This author finds the work of Koen more useful to his own style and so makes Koen's treatment of heuristics a key part of every university course he teaches (Warman and Modesitt 1988).

All of us have used heuristics in our daily personal and professional lives. In the former, we say "best guess" or "rule of thumb." In the latter, synonyms include "engineering judgement", "wild assed guess (WAG)" and "scientific WAG (SWAG)." Koen characterized heuristics as having the following properties:

1.   When they work, they save time over algorithmic solutions.
2.   They are not guaranteed to work.
3.   They may often contradict one another.
4.   They change over time.

It is the first three characteristics which differentiate, in the case of computer software, knowledge base systems (KBSs) from algorithmic solutions. [KBSs are a super set of the more commonly heard phrase, expert systems. Only a few true expert systems now exist; most are really knowledge based systems.] See Fig. 4 for one way to view the differences between the two types of solutions.

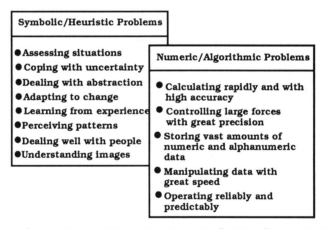

Figure 4. Conventional vs. Knowledge - Intensive Problem Characteristics.

KBSs do <u>not</u> guarantee a solution. And they certainly do not guarantee an optimal one! That simply comes with the territory -- the difficulty of the problem domain they address. See Fig. 5. This is not unique to computer-based solutions. Human beings, even domain experts, do not claim that the recommendations they give in difficult (uncertain and incomplete knowledge) situations are "guaranteed" to work![*] Others feel there is a basic distinction between "problem solving" and "decision making" -- that KBSs are good at the former and not so good at the latter. In this context, problem solving is applied to structured situations with unambiguous measures of performance. Decision making, on the other hand, applies to unstructured situations with multiple and conflicting objectives.

---

[*]In this monograph, we are obviously concerned with KBSs in the domain of civil engineering. However, should the reader be interested in exploring more about KBSs in other engineering fields, as well as having a gentle introduction to the whole field of KBSs, the classic book in expert systems by the late Donald Waterman has withstood the test of time (Waterman 1986).

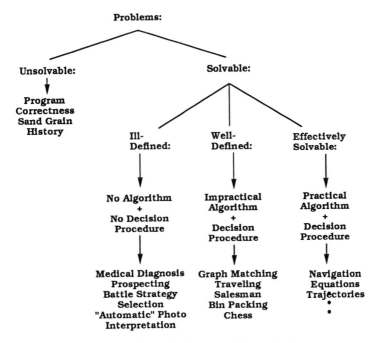

Figure 5. Domain of Expert Systems.

This is a natural segue to the fourth characteristic of heuristics as given by Koen -- that they change with time. Bridges you design today are not the same as you would have designed and constructed in Roman or Renaissance or early 20th century times. Some of the principles may remain the same, and we can recognize them all as being bridges. However, they are not the same! You have learned better (more reliable, effective, efficient, etc.) techniques. For example, the bridges of today normally have greater spans and can carry more weight. (Of course, they still fail! More about that later...)

The construction of the Sears Tower or World Trade Center a century ago would not have been possible. Civil engineers lacked the knowledge and materials available today. The state-of-the-art (SOTA) has changed.[**]

It is this fourth characteristic of heuristics -- of revealed change over time -- that is the primary subject of this monograph. Sometimes, this change arises from natural phenomena or from earlier flawed or incomplete reasoning. Other times, the change occurs as we discover better ways to translate the hidden knowledge in the mind of the expert. Later chapters are filled with structures resulting from new and improved heuristics. Heuristics change over time as a result of lessons learned and new knowledge gained.

The change does not occur in a vacuum. Most of the time, it results from the inadequacy of previous methods to meet desires and expectations. Users of engineering products normally expect that the quality of these products will naturally improve over time. The Japanese have understood this extremely well, as we see the increasing presence of their product in Western cultures. Their definition of quality is much more expansive than ours, and can be paraphrased as "the minimum cost to society after the product is shipped or service rendered" (Taguchi 1984).

One rather dramatic way in which the lack of quality is manifested is failure to perform. All engineering products have failed at times. The Tacoma Narrows bridge and Hyatt Hotel Skybridge collapsed. The Exxon Valdez supertanker released 11,000,000 gallons

---

[**]See an article by Koen for an excellent discussion of the role of the SOTA in the engineering profession (Koen 1985).

of crude oil in a pristine Alaskan bay. The external fuel tank of the space shuttle Challenger exploded. A computer virus brought a national network to a standstill.

Innovative people with engineering backgrounds also see ways that natural disasters can be mitigated. London, Chicago and San Francisco were consumed by fires. Tokyo, Managua, and Soviet Armenia lay in rubble after massive earthquakes. Johnstown, PA and the country of Bangladesh were devastated by floods.

This then is <u>why</u> we are interested in acquiring knowledge. We wish to change our heuristics over time to do a better job of filling gaps between what is currently and what could be in the future. We wish our functional needs to be satisfied in a quality manner: effectively (minimum loss to society) and efficiently (with minimum expense).

3. What is Knowledge Acquisition?

At the risk of repeating some material from Chapter 1 of this monograph on human and knowledge acquisition, it is useful to delineate various aspects of "knowledge" before discussing what it means to acquire it for ourselves or a computer program, or to elicit it from another person. In this section, we explore various ways to classify knowledge, the importance of the context in terms of <u>how</u> the knowledge will be used, and some comments on the current state-of-the-art in knowledge acquisition.

Taxonomies of Knowledge

Benjamin Bloom, a psychologist at the University of Chicago in the mid 1950's, made a major contribution in this area (Bloom, 1956). His work on a taxonomy for cognitive level skills is one this author has found useful over the course of twenty years in both academic and industrial settings. Bloom formulated six categories of cognitive knowledge. They recently appeared in a "tower" format in an article on a proposed graduate curriculum for software engineering (Ford and Gibbs, 1989). It is this format which is reproduced in Fig. 6.

As a simple example of why such a taxonomy is useful, consider the different types of knowledge being applied in the following cases:
1.   reiterating a memorized list of new German vocabulary,
2.   designing an expert system for test analysis on the space shuttle main engine,
3.   performing a study of various tradeoffs involved in designing truss members for a 50 story building in Paris, and
4.   calculating a multivariable definite integral.

To give a specific example of the usefulness of Bloom's categories, the author has included a segment from his syllabus for an introductory course on knowledge based systems (Warman and Modesitt 1988). See Fig. 7. This format, when applied consistently throughout a course, lets the student, professor, peers, administrators, and accrediting agencies know much more about student outcomes than is normally the case.***

There are, to be sure, an almost endless list of ways to classify knowledge. Waterman uses a list for the domain of knowledge based systems. See Fig. 8 for another categorization, that of various types of problem-solving tasks which have been used by knowledge engineers to classify resulting systems in a domain-independent manner. Other people, including those from instructional design, would probably refer to such a list as "task analysis" or even problem-solving categories. In the world of KBSs, it is now quite common to classify most current systems as being diagnostic in nature. Design systems are not far behind, at least in the engineering arena. Repair and planning are less common, although they are being emphasized as potential applications for Space Station Freedom (NASA 1985).

***The author has used this classification for computer science courses at all levels since 1972.

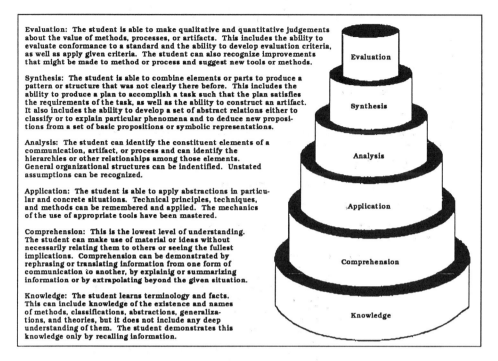

Evaluation: The student is able to make qualitative and quantitative judgements about the value of methods, processes, or artifacts. This includes the ability to evaluate conformance to a standard and the ability to develop evaluation criteria, as well as apply given criteria. The student can also recognize improvements that might be made to method or process and suggest new tools or methods.

Synthesis: The student is able to combine elements or parts to produce a pattern or structure that was not clearly there before. This includes the ability to produce a plan to accomplish a task such that the plan satisfies the requirements of the task, as well as the ability to construct an artifact. It also includes the ability to develop a set of abstract relations either to classify or to explain particular phenomena and to deduce new propositions from a set of basic propositions or symbolic representations.

Analysis: The student can identify the constituent elements of a communication, artifact, or process and can identify the hierarchies or other relationships among those elements. General organizational structures can be indentified. Unstated assumptions can be recognized.

Application: The student is able to apply abstractions in particular and concrete situations. Technical principles, techniques, and methods can be remembered and applied. The mechanics of the use of appropriate tools have been mastered.

Comprehension: This is the lowest level of understanding. The student can make use of material or ideas without necessarily relating them to others or seeing the fullest implications. Comprehension can be demonstrated by rephrasing or translating information from one form of communication to another, by explainig or summarizing information or by extrapolating beyond the given situation.

Knowledge: The student learns terminology and facts. This can include knowledge of the existence and names of methods, classifications, abstractions, generalizations, and theories, but it does not include any deep understanding of them. The student demonstrates this knowledge only by recalling information.

Figure 6. Bloom's Taxonomy of Educational Objectives.

Category	Problem Addressed
Interpretation	Inferring situation descriptions from sensor data
Prediction	Inferring likely consequences of given situations
Diagnosis	Inferring system malfunctions from observables
Design	Configuring objects under constraints
Planning	Designing actions
Monitoring	Comparing observations to expected outcomes
Debugging	Prescribing remedies for malfunctions
Repair	Executing plans to administer prescribed remedies
Instruction	Diagnosing, debugging, and repairing student behavior
Control	Governing overall system behavior

Figure 8. Knowledge Engineering Categories of Tasks.

COMPUTER SCIENCE 458/G
EXPERT KNOWLEDGE-BASED
SYSTEMS

SUBJECT MATTER

Expert (Knowledge-based) Systems

PURPOSE

To provide computer science students with the skills required to analyze, construct, evaluate, and justify expert knowledge-based systems.

PREREQUISTITES

CS 456: Artificial Intelligence, or equivalent

GOALS (Objectives)

We will use Bloom's taxonomy for the cognitive domain.
COGNITIVE

(1) knowledge: the student will be able to define, recognize, and
recall concepts related to expert systems, e.g.
knowledge, expert, knowledge system, expert system,
transparency, knowledge base, inferences (forward
and backward chaining), knowledge engineering,
knowledge acquisition (learning), expert system
shells, uncertainty measures, rules, frames, scripts,
blackboards, rehosting
the student will be aware of:
existing expert systems which use the above concepts
existing tools for building expert systems
current research issues in expert systems

(2) comprehension: the student will be able to compare and contrast major
concepts related to:
expert systems,
expert systems and "normal" software,
expert systems and artificial intelligence
knowledge engineering and software engineering,

(3) application: the student willl be able to apply methodologies and
existing tools for knowledge engineering

(4) analysis: the student will be able to analyze:
existing expert systems,
existing tools for building expert systems,
customer's need, including problem domain, and proposed
designs and testing procedures of their own and others

(5) synthesis: the student will be able to construct:
a simplified expert system using an example-based
commercial tool and demonstrate the result
an evaluation of existing commercial shells
a prototype for an expert system of her/his choice using
any available tools and demonstrate the result
a 45 minute mini-tutorial on an aspect of expert systems
for class and evaluate the result

(6) evaluation: the student will be able to evaulate various existing
expert system applications, commercial tools,
problem domain, methodologies, professional literature

AFFECTIVE　　　the student will communicate formally and informally
via oral and written means to peers, the instructor,
and, perhaps, other outside the class ("customers")
the student will actively participate as a member of
a variety of different teams, e.g., user, expert,
developer, instructor

Figure 7. Example of Bloom's Cognitive Levels in a Syllabus for a Course on Knowledge-Based Systems.

Taxonomies such as the above ones should be exceptionally useful when investigating how to learn different types of knowledge. Knowledge acquisition is, after all, the main theme of this entire book. Surely, different methods come into play when a student is learning rudimentary verb ending forms in a second language than when she is learning to evaluate several requests for proposals for a space station power system! Yet, paradoxically, there has been a dearth of written work in the computer field which even begins to relate work on learning with what type of knowledge is being learned. One suspects it is only because one of the authors has her Ph.D. in educational psychology (McGraw) does the issue even appear in a recent book (McGraw and Harbison-Briggs 1989). See Fig. 9 for one way in which they have typed learning with various modes of knowledge. We will use this figure as the basis for structuring our discussion of knowledge acquisition techniques in the section on "Heuristic Cookbook for Knowledge Acquisition with the Computer as Consumer."

According to McGraw and Harbison-Briggs, procedural knowledge includes the skills which an individual knows how to perform, such as language or motor skills. Declarative knowledge represents surface-level knowledge which people can verbalize. Semantic knowledge is a type of long-term memory involving organized knowledge containing facts, relationships, concepts, etc. Episodic knowledge contains information about episodes grouped by time or space.

The major exceptions to such a dearth are from the field of psychology, particularly at the level of children (Piaget 1950 and Papert 1980) and from the field of instructional design (Harmon and King 1985). Seymour Papert at MIT has directly addressed the problem-solving levels of Bloom (analysis, synthesis, and evaluation) with the creation of his LOGO language which is so highly-touted in preschools and elementary schools for "self-discovery." Harmon uses his background in industry training to point out the impact time has on our learning both experiential and causal types of knowledge. The former is amenable to heuristics, and the latter more explainable with the use of a deep model based on theory. Researchers in learning have a big responsibility to fill this void of knowledge taxonomies. What type of knowledge (according to Bloom or anyone else) is being acquired by the learning agent?

KNOWLEDGE	ACTIVITY	SUGGESTED TECHNIQUE
Declarative Knowledge	Identifying general (conscious) heuristics	Interviews
Procedural Knowledge	Identifying routine procedures/tasks	Struct. Interview Process Tracing Simulations
Semantic Knowledge	Identifying major concepts/vocabulary	Repertory Grid Concept Sorting
Semantic Knowledge	Identifying decision making procedures and heuristics (unconscious)	Task Analysis Process Tracing
Episodic Knowledge	Identifying analogical problem solving heuristics	Simulations Process Tracing

Figure 9. Correlation between Knowledge Type and Acquisition Technique. (Karen L. McGraw/Karan Harbison-Briggs, KNOWLEDGE ACQUISITION: Principles and Guidelines, (C) 1989, p. 23. Reprinted by permission of Prentice Hall, Inc., Englewood Cliffs, New Jersey.)

Given all the above, what then is knowledge acquisition, as currently understood? McGraw, borrowing from Hayes-Roth (Hayes-Roth 1983) states that knowledge acquisition refers to " 'the translation and transformation of problem solving expertise' from a knowledge source (e.g., human expert, documents) to a program." Because we are trying to define and understand the concept of knowledge acquisition, we will consider the more general case where the destination can be a human, or any type of system.

Let us then first investigate the most powerful, enduring, and useful methods for knowledge acquisition in general. Then, various filters for knowledge could be applied in light of current state-of-the-art techniques in translating the knowledge to a "program." See Fig. 10. However, recall Koen's admonition about heuristics -- they change over time! Far better to remind the reader of successful knowledge acquisition methods in general. Some of you may well be the ones to augment the current state-of-the-art -- to add to the current (barren) storehouse of those methods which are applicable to computer programs.

Context of Learning

In a graduate course on artificial intelligence at Carnegie Tech nearly 25 years ago, the author listed to Saul Amarel, then of RCA, lecture on learning. A point Amarel proposed then made eminent sense. Paraphrased, it was "it is not enough to get a machine to learn, we must understand how we can make a machine learn to learn." The distinction between "learning" and learning to learn" is a key one. For not only must we be aware of the knowledge taxonomy, as discussed in a recent section, but we must also concern ourselves with how the resulting knowledge, once learned, will be used. Will it be used to perform some task, or will it be learning for the sake of learning? There are two major organizations which address one or the other -- those of training and those of education.

Currently, those of us working in knowledge based systems concentrate on the "training" side, for the most part. We wish to persuade, cajole, program some entity to learn to perform some task. However, recalling Amarel's phrase, there is a clear interest in having computers become involved as "recipients" on the "education" side of the picture. For the purposes of this monograph, and the current state-of-the-art, we will also concentrate on the "training" side of the diagram. But the time is coming when we will devote more efforts to the "education" side. Several of you may already be aware of the growing interest in an area known as intelligent computer assisted instruction or intelligent tutoring systems which help students learn (Mandl and Lesgold 1989).

State-of-the-art in Knowledge Acquisition

Let us take a little snapshot of where we currently are with regard to what we commonly recognize as involving knowledge acquisition, and some of the more commonly used terms. See Fig. 11.

This is only a partial snapshot, yet the immense amount of experience stretching over literally thousands of years which involve the acquisition of knowledge by an agent, normally a human being, is obvious. Clearly, we have vast resources. The Roman Empire did, after all, know something about building bridges and roads, as did the Chinese about walls Egyptians about pyramids! Consequently, it is a major source of puzzlement to the author as to why most current research in knowledge based systems which involves the knowledge acquisition component virtually ignores the wealth of resources revealed in the above figure.

Instead, we invent new (high-priced!) terms such as "knowledge engineer!" Only a few people have recently indicated publicly that the years of instructional design experience might actually be useful (NTU 1989). Yet industry, for years, has made use of instructional design for many tasks involving the transfer of expert knowledge to human students. Journals addressing this audience abound, such as those of the National Society for Performance and Instruction and Educational Technology. The trained employees are then expected to perform the newly-learned task with more competence. It does not take a genius to see the similarity with the construction of expert systems. Researchers and practitioners alike in knowledge acquisition would do well to make some good friends in the professional training community.

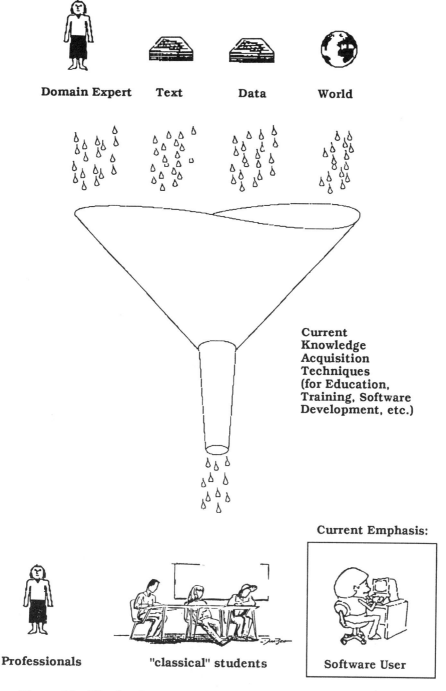

Figure 10. Filtering from a Knowledge Source to a Destination.

CONTEXT	SETTING	KNOWLEDGE SOURCE	RECIPIENT	POSSIBLE COMPONENT OF PROCESS	COMMON NAME
Education	classroom	professor	student	textbook	"Classic" education
Education	computer labs	computer	student	multi-media	computer-based education "CBE"
Education	computer labs	expert+comp. (domain + professor)	student	multi-media	intelligent CBE "ICAI"
Training	work	supervisor	employee	none	On the job training "OJT"
Training	work	computer	employee	multi-media	computer-based train "CBT"
Training	work	human expert (domain)	employee	instructional design + trainer	"classic" training
Training	work	list of directions	employee	none	job perfor-mance aid "JPA"
Performance	work	human expert (domain)	computer	knowledge engineer	expert system
Performance	work	human expert (domain)	computer	engineer	software

Figure 11. Partial Look at the State - of - the - Art in Knowledge Acquisition.

4. How is Knowledge Acquisition Performed?

Now we come to a key section of this chapter. Earlier, we looked at <u>why</u> we are interested in knowledge acquisition -- to change our heuristics over time so that we can do a better job of solving significant problems. We want to improve our abilities to fill the gaps between what current situations are, and what futures we would like to see -- the gaps between "what is" and "what should be." Then, we have just finished a section on <u>what</u> knowledge acquisition is. There, we argued that the definition normally given is too narrow: "the translation and transformation of problem-solving expertise from a knowledge source to a program." The recipient should be expanded to <u>any</u> entity, and then filtered down to the level currently understood by a computer program only at the last step.

We will first look at the case where <u>humans</u> are the recipients of knowledge acquisition in a <u>structured context</u>, splitting the cases into past, present, and future. We do this in hopes that the reader may some day apply these successful methods to cases where a computer program is the recipient. Only a short discussion of each is given, as the reader is undoubtedly familiar with most. However, seeing them in this context may spark her/him to have some new insights about how to apply the techniques to helping "programs" to learn. We also show how critical the passage of time is to leaning -- expert people <u>and</u> expert systems acquire expert knowledge gradually, not overnight.

Before starting, however, please recall that the presence of a new tool or technique does <u>not</u> automatically render the former ones obsolete. Clearly, the current methods for helping humans acquire knowledge use the former methods of classical education and on-the-job training. The following parable of the '67 Chevy is instructive in this context. See Fig. 12.

ONCE UPON A TIME THERE WAS A MAN WHO WANTED TO BE ABLE TO GET FROM NEW YORK TO LOS ANGELES ON 12 HOURS' NOTICE. HIS MEANS OF TRANSPORTATION WAS A 1967 CHEVROLET THAT HAD BALD TIRES AND FIRED ON THREE CYLINDERS.

RECOGNIZING THE DEFICIENCIES OF HIS MEANS, HE HAD THE CAR TUNED UP AND BOUGHT NEW TIRES, A SUPERCHARGER AND A RADAR DETECTOR. PERFORMANCE IMPROVED 87%, BUT HE STILL COULDN'T GET FROM NEW YORK TO L.A. IN 12 HOURS.

HE NEXT PUT IN A CADILLAC ENGINE AND ADDED STREAMLINING: STILL NOT FAST ENOUGH. HIS LAST GASP WAS TO INSTALL A TURBINE ENGINE AND AERODYNAMIC CONTROLS, WHICH GOT HIM TO 200 M.P.H. IN THE INTERSTATES - WITHIN REACH OF WHAT HE NEEDED - BUT LED TO BAD SCENES GOING THROUGH THE SMALL TOWNS.

WHEN LAST HEARD OF, HE WAS COMPLAINING ABOUT PROBLEMS WITH HIS TURBINE ENGINE AND SMALL TOWN POLICE; HE HAD NOT REALIZED THAT HE WAS TRYING TO IMPROVE THE PERFORMANCE OF A BASICALLY WRONG MEANS OF TRANSPORTATION.

HE WAS ALSO ASKING WHAT CAUSED CONTRAILS AND, WHEN TOLD ABOUT JET AIRPLANES, SAID, "THEY'LL NEVER FLY!"

MORALS:

● IF YOU'RE USING THE WRONG TOOL, IMPROVING IT'S PERFORMANCE WON'T HELP MUCH.

● THE PEOPLE WHO KNOW THE MOST ABOUT EXISTING TOOLS ARE OFTEN THE HARDEST TO GET TO ACCEPT BETTER AND DIFFERENT ONES.

● THE DEVELOPMENT OF NEW TOOLS SELDOM MAKES THE EXISTING ONES TOTALLY OBSOLETE. EVEN IF OUR MAN ACCEPTED JET AIRPLANES, HE WOULD STILL HAVE NEEDED HIS CAR FOR GETTING TO THE GROCERY STORE.

Figure 12. The Parable of The '67 Chevy.

Humans as Acquirers of Knowledge

Human beings have discovered a wide variety of methods for the transmission of knowledge from one to another. Long before books or written symbols appeared, the oral tradition was well established, In fact, were a caveman to walk into many classical classrooms today, whether in an academic or industrial setting, he would be very comfortable with the primary mode of delivery, even if not the content.

An abbreviated history of delivery modes used in learning is given in Fig. 13 (modified from Modesitt 1983).
Several points bear elaboration.
1. The first textbook was one filled with graphics -- authors of current ones should take note.
2. Humans use tools extensively in the learning process -- from books to computers to satellites.
3. Most delivery modes, after Socrates, concentrated on increasing the <u>efficiency</u> of instruction, i.e., more students/teacher.
4. Only relatively recently has increased <u>quality</u> become a concern -- of maximizing the learning of an individual. Mostly this has been accomplished by an emphasis on learning by doing and interactive learning. Hence the use of computer based learning (CBL) is continually growing and permeating more and more of our learning institutions. We seem to have a dearth of well-qualified teachers, so a "surrogate" in the Socratic form of interactive CBL is an alternative.

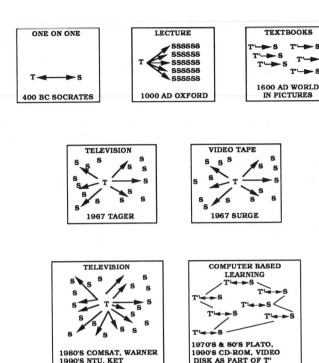

Figure 13. Delivery Modes of Learning (T=Teacher,
S=Student, T'=Teacher's Aid).

For those of you who are dubious about the presence of computer-related technology in learning environments, go browse through your child's school or your corporate training facility. Times have changed, and have done so extraordinarily quickly! For example, Kentucky, a state not known as a rich state, is now the first one in the nation to have a satellite down-link in every single K-12 school (Welch 1989). Each school also has computers connected in a backward link via telephone circuits.

Large national and international organizations are keenly aware of the dramatic upsurge which technology is playing in the process of human acquiring knowledge. ASEE and IEEE have long sponsored such conferences. The Association for the Development of Computer-based Instructional Systems (ADCIS) has a history of over thirty years, only a dozen years younger than the Association of Computing Machinery (ACM), the "grandfather" association for computing. Apple Computer Corporation was founded in 1975 and established itself in schools nationwide. Today, the gross annual sales exceeds $5 billion. National Technological University (NTU), a consortium of about 20 major engineering universities, just celebrated its fifth year of delivering accredited M.S. degrees in seven engineering fields via satellite to industry sites (NTU 1990).

What of the future? As mentioned earlier, most current methods will still be useful in the years and generations ahead. It is reasonable to predict increasing penetration of the technologies. However, it is dubious that these will be the only ways in which humans will acquire knowledge in 2010, 2050 or 2100 A.D.

The Significance of Time in Acquiring Knowledge for Human Beings

The following simple observation has considerable impact: Knowledge Acquisition is a lifelong process!
Humans do not progress in knowledge from being completely ignorant one day to being an expert by the next day. In the current context, human experts become experts over decades of new experiences. They require years upon years to make the following transformation:

Data --> Information --> Knowledge --> Wisdom

These stages are just as true for our personal lives: marriage, children, etc., as they are for acquiring knowledge in our respective professions.

Moreover, experts maintain their expertise by continuing to practice it in dialogues with others. To proclaim a person as an expert and then to "jail" them, prohibiting any further growth or contact is ridiculous. Fig. 14 is taken from a paper containing some admonitions about doing precisely this with human experts and expert systems (Modesitt 1987).

Figure 14. Locking up the Expert.

The author recently heard the following statement: "90% of the world's knowledge has been generated in the last 30 years." Since the author graduated from high school over 30 years ago, the phrase was striking. Without speculating unduly, "knowledge" probably meant "words or pages in journals, etc." In any case, how much increase in <u>wisdom</u> has occurred in the last 30 years -- 90%?

Paul Harmon, an instructional technologist for over 20 years (and a recent convert to knowledge based systems), has constructed a useful graph depicting how the knowledge sources change over time for humans (Harmon and King, 1985). See Fig. 15.

Humans start life learning from their parents, siblings, environment, and pre-school friends. More structured learning occurs as they move into educational institutions such as K-12 and the university, at least for some. After they leave school, most of what they learn comes from experience (day-to-day events, friends, mentors, successes and failures). To be sure, those of us who are professionals continue with some knowledge acquisition in a structured context: workshops, training classes, continuing education via the Association for Media-based Continuing Education for Engineers (AMCEE) and the National Technology University (NTU) in the United States. There are also hundreds of companies and thousands of consultants who address just this need for continuing education.

Note well, however, that <u>most</u> of what a human learns comes from daily experiences, <u>not</u> from reading books or journals, or sitting in a classroom. Oral and visual learning are fine, but most humans are kinesthetic.

"To hear is to forget,
To see is to remember,
To do is to learn."

supposed Chinese proverb

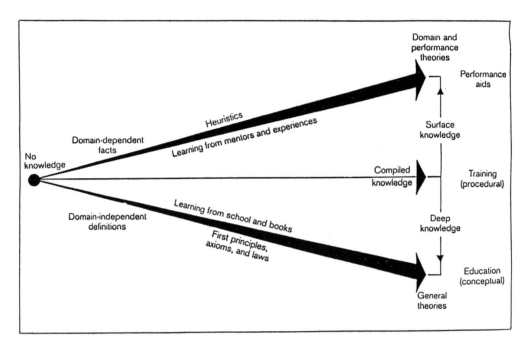

Figure 15. Change in Knowledge Sources for Humans over Time.
(Paul Harmon/David King, EXPERT SYSTEMS: Artificial Intelligence in
Business, (C) 1985, p. 33. Reprinted by permission of John Wiley & Sons,
Inc., New York, NY.)

We would do well to emblazon this truism on our minds when we speak of knowledge acquisition in the context of computers learning.
  - <u>How</u> can we get a computer to "experience?"
  - <u>Can</u> we get a computer to "experience?"
  - <u>What</u> can we get a computer to "experience?"
  - <u>Should</u> we get a computer to "experience?"
  - Are we trying to build the analog of an airplane
    which flaps its wings, just like we observe birds doing?
Some of these questions will be addressed in the next section, but no answers are given. Having a computer "experience life" is an alien concept to human beings. It will probably remain so for the lifetimes of all of us, and those of our grandchildren, at least.

Computers as Acquirers of Knowledge

Lest you wonder why we are asked to ponder the previous questions about how human beings acquire knowledge, please recall that heuristics change over time. You, and particularly the younger among you, will be asked to <u>design and build</u> computer based systems which learn! Moreover, you will also be requested to investigate the application of your knowledge (heuristics) to construct systems which learn to learn.

The desire for Space Station Freedom to permit an autonomous operational and maintenance mode is well documented (NASA 1985, Broad 1990). Just how do you suppose this structure will be built? See Fig. 16. Will all of you, civil and structural engineers, go up in the shuttles to determine the environment and oversee construction? Will you send up hordes of human construction workers? Will you give the task specifications and performance requirements to the robot builders? If you won't, who will? Do you wonder about your cohorts in Japan who are looking 20 years ahead for construction of large structures in space? Will we be working with them, now that they also have shown their ability to orbit a satellite around the Moon, courtesy of an engine built by Nissan Corporation (Sanger 1990)? What form of response will we have to the request of the Soviet Union to engage in a cooperative venture to explore space and to construct habitats both in space and on other bodies (Schmitt 1986)? How about the lunar and Martian bases which are sure to come, as part of the newly-announced space exploration initiative (SEI) of the current administration?

With the case of a human being as a knowledge acquirer well in mind, let us move on. What are the analogs for the instances wherein we wish a computer (including both hardware and software) to embody knowledge? Again, it is useful to survey this question from the viewpoint of time: past, present, and future. Just as was the case for human learners, it is critical to remember that the items under the later headings, e.g., "future," should also be augmented by items under previous ones. For example, software engineering is a technique useful now, as well as in the future.

Figure 16. Proposed Single Keel Configuration for Space Station Freedom.

1. Past Methods

Most of the knowledge communicated to computers in the past has been that which was "programmed in" via explicit procedures. The taxonomy of Bloom, introduced in an earlier section, is helpful here. The 100,000,000,000 lines of code of COBOL and virtually all of the existing Fortran code represents knowledge of the first three of Blooms levels: memorization, comprehension, and application. Computing flow rates of liquid oxygen (LOX) in a space shuttle main engine is one example. Determining your net monthly salary is another. So is looking up an article in an ERIC automated catalog.

The field of artificial intelligence (AI) has concentrated on communicating knowledge to computers at the higher levels of Bloom's taxonomy: analysis, synthesis, and evaluation. That this effort has not been 100% successful is obvious by the checkered past of the AI field: many predictions of short-range successes in the 1960's by "experts" are still far from being attained.

During the early 1960's, the field of AI was split into four (at least!) major camps: those of the cognitive, algorithmic, physiological and robotic schools.

Cognitive
> Interested in how human beings think at a symbolic level and how that thinking can be represents in a machine, e.g., protocol analysis.

Algorithmic
> Interested in solving very difficult symbolic tasks, e.g., theorem-proving and integration in the calculus, regardless of any resemblance to human thought processes.

Physiological
> Interested in how the human brain works at the anatomical level, e.g., perceptrons, neurons, synapses, threshold logic.

Robotics
> Interested in simulating human abilities of vision, touch, motion. Often coupled with models from cognitive area.

Clearly, this is a miniscule summary. The reader is referred to all three volumes of the recently-updated Handbook of Artificial Intelligence for a detailed treatment (Feigenbaum and Barr 1989).

2. Current methods

The most obvious current method is that of the knowledge based system (KBS). This monograph is replete with such examples, so we need not discuss these "acquirers of knowledge" any further, other than to point out that KBSs are only one application of AI principles.

The robotic school, mentioned as a component in the past, is still alive and well in the present, particularly in Japan. Robots are well represented on assembly lines. They are under active study at JPL (again) as a means of exploring lunar and Martian surfaces. And they are serving as security guards at some locations.

Although not as popular as KBSs, the dominant mode today for communicating knowledge to computers is via software engineering products. Production of software is a $250,000,000 business in 1990, and is expected to grow to become $400,000,000 by the 2000 A.D., easily surpassing GM and Exxon. These software products, ideally, solve real needs of customers in a cost-effective manner. They are large scale complex programs for solving problems at the lower levels of Bloom's taxonomy. Examples include FAA flight control systems, fly-by-wire aircraft systems, space shuttle operations, and control of manufacturing and chemical process lines.

3. Future methods

Again, the author lays no claim to possessing crystal ball powers. R2D2 and HAL 9000 are not however just around the corner! There are a few major developing themes: neural networks, embedded systems, and integration of KBS and software engineering.

Neural nets are attracting considerable attention today, primarily for classification tasks. As outgrowths of the perceptrons of nearly thirty years ago, they still embody much physiological jargon. They are useful for certain domains, e.g., pattern recognition, particularly when hosted on some of the new massively parallel hardware such as hypercube and connection machines.

Embedded systems are a form of KBS system in which the intelligence is actually part of the physical instruments or entity. Automatic braking systems are primitive examples. Robots capable of self-repair would be much more advanced examples.

The most promise however lies in the successful integration of software engineering and KBSs (Modesitt 1989). The former brings the knowhow of constructing and maintaining large scale useful software systems on time and within budget. The latter brings an increase in the allowable complexity and transparency of the system to the end user. It will also begin to let users trust their systems more and will more clearly show where and why the same systems fail. In a very real way, when systems are able to explain, in a meaningful way, their answers and/or their questions, we are then justified in indicating that they "know" the subject at hand. We have highly respected company in holding such a view. See Fig. 17. There is a growing awareness of just how critical such an explanation-based interface is for acceptance by users. Donald Michie, "grandfather" of artificial intelligence in the United Kingdom, is adamant about such an interface (Michie 1987). Consequently, an entire subfield of explanation-based learning is now making its presence felt (Ellman 1989).

... WHEN, THEREFORE, ANYONE FORMS THE TRUE OPINION OF ANYTHING WITHOUT RATIONAL EXPLANATION, YOU MAY SAY THAT HIS MIND IS TRULY EXERCISED, BUT HAS NO KNOWLEDGE; FOR HE WHO CANNOT GIVE AND RECEIVE A REASON FOR A THING, HAS NO KNOWLEDGE OF THAT THING; BUT WHEN HE ADDS RATIONAL EXPLANATION, THEN HE IS PERFECTED IN KNOWLEDGE ...

... TRUE OPINION, COMBINED WITH DEFINITION AND RATIONAL EXPLANATION, IS KNOWLEDGE ...

Theatetus, in the PHILOSOPHY OF PLATO
Jowett Translation, 1927, p.564.

Figure 17. The Importance of Explanation Capabilities.

The author has written at length on the importance of integrating knowledge-based systems and software engineering in a real world industry environment. The most recent such publication (Modesitt 1990a) also integrates computer-based learning as well. Scotty, a software system for performing test analysis on space shuttle main engines, was built over four years using both KBS and software engineering methods and tools. These tools included an expert system building tool (ESBT) and a computer aided software engineering (CASE) one. The former generated Fortran code, which was therefore easily understandable and maintainable by the engineering users, mechanical and aerospace ones in this context. The CASE tool for documenting software development satisfied the software quality assurance group of both our NASA customer at Marshall Space Flight Center, and the internal Rockwell organization. Moreover, the CASE tool was extremely easy and natural to use for the engineers.

There is a growing, but still small, awareness on the part of the KBS and software engineering professional communities that they have much to gain from one another (Simon 1986). A number of joint conferences are being held (Rich and Waters 1986). A journal has just appeared entitled the International Journal of Software Engineering and Knowledge Engineering.

One of the most powerful ways in which this interaction could occur is at the "front-end" of the requirements phase. This is where a user and the developer are mutually trying to understand what the user desires. For a sizable number of applications, the use of an inductive tool is a very welcome addition for this task. In fact, this author fully anticipates that some CASE vendor will announce shortly that the code generation component of their tools will use the powerful inductive techniques now commonplace in many ESBTs. This same technique of generating code from an unordered list of examples compiled by an expert or from a historical data base of records was used for Scotty. Data indicate a ten-fold increase in code generation productivity, on the order of 100 lines-of-code/day (Modesitt 1990b, 1988). The topic of induction will be covered more fully in the next chapter. However, it should be pointed out that the topic is even now appearing in introductory AI textbooks, in both manual and automated forms (Patterson 1990).

In summary then, we have learned a very great deal in a relatively short period of time about how to help computers acquire knowledge. We have been able to have computers perform Bloom's level one type of knowledge, that of memorization and recall, for over forty years. The higher stages of analysis and synthesis have become realities in the market only within the decade of the 1980s. And we would do well to remember the normal latency period of 15-20 years for an idea to migrate from a fuzzy idea in some research environment to a success in an every-day engineering one.

4. Stages in Knowledge Acquisition for Computer Programs

Recall the earlier discussion of how humans progress in learning over time. We draw from different knowledge sources during our lifetime. We move from our parents to textbooks to unstructured successes and failures of adult life, both professionally and personally.

Thus, it is not surprising that some people have acknowledged the importance of stages or phases in the progress of a computer acquiring knowledge. Some, but not much, of the subfield of machine learning is concerned with just this question. More often, the illegitimate assumption is made that the memory of a computer goes from a tabula rasa format (blank sheet) one moment, to that of an acknowledged expert the next.

DARPA acknowledges the phases of computer learning in the following way (Merrifield 1986). See Fig. 18.

Aide:	automated control of a single mission operations sub-system
Apprentice:	automated control of multiple subsystems
Assistant:	hierarchical control of multiple subsystems
Associate:	distributed control of multiple subsystems

Figure 18. Phases of Learning for Pilot's Associate of DARPA.

Skill Level	Decision	Commitment
Novice	Analytical	Detached
Advanced Beginner	Analytical	Detached
Competent	Analytical	Detached understanding and deciding. Involved in outcome
Proficient	Analytical	Involved understanding Detached deciding
Expert	Intuitive	Involved

Figure 19. Stages of Skill Acquisiton according to Dreyfus.

Hubert Dreyfus, a contemporary philosopher and long considered by some to be a gadfly in the marvelous ointment of artificial intelligence, has formulated another set of stages (Dreyfus and Dreyfus 1986). He and his applied mathematician brother argue, with some success, that humans go through the stages given in Fig. 19.The remainder of their book is a detailed set of reasons for why computers will <u>never</u> progress from "competency" to "expert." Although controversial, Dreyfus provides a much needed "other side" to the optimistic euphoria which usually permeates the field of artificial intelligence.

5. Heuristic Cookbook for Knowledge Acquisition with The Computer as Consumer

The methods for knowledge acquisition which follow are indeed heuristic in nature. They do <u>not</u> guarantee a solution to your effort of soliciting knowledge in order to solve a major engineering problem. They may fail and they sometimes contradict one another. But they <u>are</u> the best currently in use. That will probably <u>not</u> be the case tomorrow. <u>You</u> may come up with improved versions!

The author is indebted to Dr. Karen McGraw for many parts of this heuristic cookbook. Her seminal work in conjunction with Karen Harbison-Briggs made the extensive folklore of knowledge acquisition accessible to most of us for the first time in a useful format (McGraw and Harbison-Briggs 1989). Fig. 20, due mostly to McGraw, shows what we intend to provide.

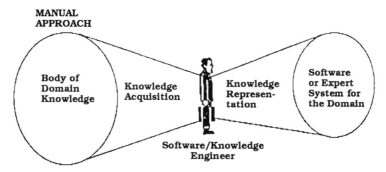

Figure 20. Transformation of Knowledge from Source to KBS, via Human Intermediary. (Karen L. McGraw/Karan Harbison-Briggs, KNOWLEDGE ACQUISITION: Principles and Guidelines, (C) 1989, p. 276. Reprinted by permission of Prentice Hall, Inc., Englewood Cliffs, New Jersey.)

The approach discussed in the next chapter by Arciszewski and Ziarko has the same source and same destination, but a different intermediary. See Fig. 21.

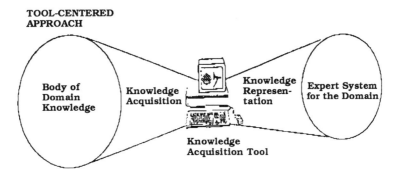

Figure 21. Transformation of Knowledge from Source to KBS, via Computer Intermediary. (Karen L. McGraw/Karan Harbison-Briggs, KNOWLEDGE ACQUISITION: Principles and Guidelines, (C) 1989, p. 276. Reprinted by permission of Prentice Hall, Inc., Englewood Cliffs, New Jersey.)

Knowledge Acquisition as a Process

Knowledge acquisition, like many other tasks, can be decomposed into a "program" or process involving the flow of data and control. A flowchart format of knowledge acquisition is given by McGraw. See Fig. 22.

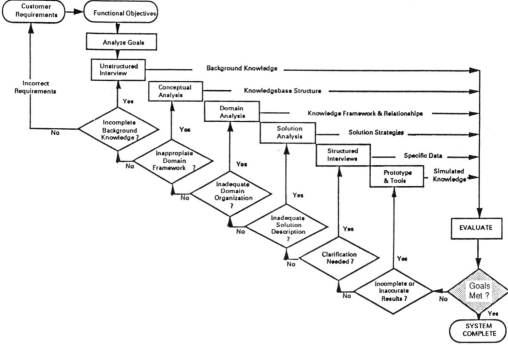

Figure 22.   Flowchart Format of Knowledge Acquisition.
(Karen L. McGraw/Karan Harbison-Briggs, KNOWLEDGE ACQUISITION: Principles and Guidelines, (C) 1989, p. 54. Reprinted by permission of Prentice Hall, Inc., Englewood Cliffs, New Jersey.)

A short verbal description of the iterative process follows. The initial task in knowledge acquisition is to identify the important knowledge required by the expert and the knowledge engineer. An unstructured interview is good for this. Afterwards, the knowledge engineer uses conceptual analysis techniques to understand and represent the way in which the expert has organized the concepts. Domain analysis is then used to determine the basic framework and relationships. Structured interviews give specific declarative data to planned questions. Solution strategies of the expert are then analyzed. Finally, automated tools are used to develop a prototype.

Catalog of Available Knowledge Acquisition Techniques

The array of knowledge acquisition techniques is vast and is drawn from a wide variety of fields, as intimated earlier. Some of these include lecture (from education and training), personal constructs (psychology), brainstorming (business), consensus (communications), task analysis (instructional design), protocol analysis (computer science), and prototyping (software engineering). Clearly we cannot give a fair or even cursory treatment to an exhaustive list of such techniques. We defer primarily to the definitive work of (McGraw and Harbison-Briggs, 1989). However, we can provide a structure for discussing six common techniques which are in current use, by referring to Fig. 9 which classifies suggested techniques according to knowledge type and activity.

1. Interviews [Activity: Identify general and conscious heuristics]

Early KBS development was characterized by a "research" mentality in which development time was relatively inconsequential. The individuals were much more interested in developing tools and a "proof of existence." Times have changed indeed in the last 20 years!

During these early years, the dominant mode of acquiring knowledge from domain experts was the interview. The task was, by the standards of today, relatively simple: diagnosis of a single task done by a single expert. Moreover, the people were normally "early adopters," and hence more than eager to devote inordinate amounts of time and energy to building a successful system. Remember, please, not to be too hasty in criticizing these individuals, when you recall the absolutely primitive nature of the tools they had available.

Today, however, with thousands of KBSs is use, and with considerably more advanced tools, much more attention is being paid to a systems approach to development, operation, and maintenance of a high quality software system. Thus, working with domain experts in a timely fashion becomes a critical issue. Hence, the interview as a knowledge acquisition techniques is being re-examined.

a. Unstructured Interview

In this oldest of the knowledge acquisition techniques, the expert assumes the role of teacher, lecturing to her student, the knowledge engineer. The latter takes notes and asks questions in a generally undirected manner. As a way to explore a problem area initially, it is a very common method. Potential problems include poor use of the expert's time and extreme dependence upon his/her ability to articulate problem solution strategies. This is the well-known knowledge acquisition bottleneck: the more expert a person is, the more difficulty he/she has in elucidating his/her heuristics.

McGraw and Harbison-Briggs, p. 187, list six problems with this type of interview:

I.   knowledge engineer and domain expert must both actively prepare.
II.   domain expert finds it very difficult to express some of the critical elements.
III.   domain expert may interpret the lack of structure as requiring little preparation.
IV.   data acquired is often unrelated and difficult to review and integrate.
V.   specific information is not acquired.

b. Structured Interview

This is a primary mode of knowledge acquisition, and one in which the knowledge engineer outlines specific sample goals and questions. This goal-orientation reduces the interpretation problems and domain expert subjectivity common to the unstructured interview.

Some steps to be followed are listed by McGraw and Harbison-Briggs, p. 188:

I.   knowledge engineer studies available source material.
II.   knowledge engineer identifies major gaps in the knowledge base.
III.   a knowledge acquisition form is designed and used.
IV.   sample questions are constructed.
V.   domain expert encouraged to prepare.

2. Process Tracing [Activity: identify routine procedures/task]

It is clearly not sufficient for a knowledge engineer to identify facts and concepts known to the domain expert. It is imperative that the knowledge engineer be able to interpret correctly the decisions regarding how such knowledge is actually used to solve a problem; both common and difficult ones. The application of factual knowledge for solving problems is of vital importance.

Process tracing is any set of techniques which permits a knowledge engineer to determine _how_ an expert solves a task or makes a decision. The roots of this include some early work by Newell (Newell, 1972) at Carnegie Tech on recording the observations of how students would solve crypto-arithmetic problems. For example, a student is asked to determine a mapping of decimal digits to letters which makes this sum true:

S E N D

+ M O R E

M O N E Y

Newell's work in this protocol analysis method is the genesis of much research in artificial intelligence even today.

Notice that this knowledge acquisition technique does not use the dialogue seen in interviews. In process tracing, the knowledge engineer _records_ the actions of the domain expert when the latter is working through a problem. The actions can either be accompanied by "thinking out loud" (as used by Newell) or be recorded for later playback and review.

This set of techniques for presenting a scenario and then asking the domain expert to solve it can be used with both routine (to the domain expert) tasks, and difficult tasks. In the latter instance, the expert is often asked to make use of analogies. Here, the interest is determining what episodic knowledge the domain expert is tapping -- what cues from previous similar situations are being used.

3. Conceptual Analysis [Activity: identify major concepts/vocabulary]

What classification and categorization techniques are used by the domain expert? Once the knowledge engineer is familiar with the domain [a BIG caveat, by the way], she can present sample classifications problems to the domain expert. The result is usually the view of a single expert, and hence limited. It does, however, yield information on both concept definition (the concepts used by the expert) as well as concept interrelationships (how these concepts are organized internally to the expert). Both of these are used heavily in other knowledge acquisition techniques to plan knowledge based structure and determine gaps.

Concept definition techniques are relatively easily grasped. They usually use a concept dictionary approach familiar to many psychologists. Classification of items into several categories is a task we all have done since childhood. For example the "marvelous" charts in high school biology listing kingdom, phylum, genus, species, sub-species, etc. are easily remembered. It is no surprise that domain experts have constructed similar ways to categorize objects of professional interest. This may be a basis of our capacity to infer relationships.

McGraw and Harbison-Briggs, p. 141, give five major techniques for identifying relationships among categories:

a. Taxonomies

This is exemplified by the high school biology reference above. They are most useful in declarative knowledge such as is frequently found in diagnostic KBSs. Inheritance of properties from members "higher" in the classification is a key feature. Object-oriented programming makes maximum utilization of this inheritance, and hence is often chosen for implementation in such domains.

b. Concept Sorting

Here, a primitive set of basic components prepared by the knowledge engineer from textual sources is presented to the domain expert. It is the job of the expert to sort the major concepts into a tree of related ones, usually interactively. This can be done manually, or partially automated with a computer.

c. Scaling

Hierarchical cluster analysis is one such method and has been a mainstay, even in fields such as document retrieval. This "closeness" metric is of use in building maximally related clusters. Psychologists have long used this method to determine degrees of closeness between concepts.

d. Conceptual Clustering

This is a set of very widely-used techniques which generates classifications from initially unclassified data. An abstraction of the domain is one result of such clustering, and is a key ingredient of inductive systems (Michalski 1983).

McGraw and Harbison-Briggs, p. 146, provide this framework: "Given a certain number of concepts, devise a classification scheme to group the objects into a number of classes (with the following constraints).
  (a)   Each of the concepts is described by a set of numerical measures (and if inductive tools are used, qualitative values, as well).
  (b)   objects within classes or groups must be similar in some respect.
  (c)   objects within classes or groups must be unlike those from other classes."

e. Repertory Grids

This technique has also been used in a host of domains, from market research to counseling. The intent is for the domain expert to represent a set of basic concepts in a specific domain, by measuring similarities and distances among objects and to represent these graphically on a grid.

A set of constructs, drawing from the personal construct theory of a psychologist (Kelly 1955), or bi-polar characteristics, e.g., "hot" and "cold", is developed. The knowledge engineer then builds sample elements which have attributes to be rated on these constructs e.g., from the "cold" of liquid oxygen to the "heat" of boiling water. The domain expert is asked to rate each element on each construct, via a Likert or Thurstone attitude scale.

The results are than analyzed via factor or cluster analysis to determine similarities and differences. This knowledge acquisition technique lends itself to computer augmentation, e.g., a spread-sheet application.

4. Task Analysis [Activity: Identify unconscious decision-making procedures and heuristics]

As the expertise of the domain expert increases, he/she becomes less and less likely to be able to articulate the methods used to solve difficult problems, i.e., the knowledge articulation bottleneck becomes even narrower. Hence, techniques such as task analysis are important as they permit a way to get to this unconscious level. The knowledge engineer is provided with methods for isolating major tasks, constraints, prerequisites, and actions used by the domain expert when solving typical problems.

This tool describes the functions performed by a human expert and also determines the relation of each task to the overall job. Once again, our friends in the instructional design and industrial training environments have been using this technique for decades.

5. Simulations and Prototype [Activity: Identify problem solving heuristics by use of analogies]

These techniques are most useful in the latter phases of knowledge acquisition. Here, they permit the domain expert to interact directly with an automated tool to solve a problem. They can finally begin to see a skeleton of the to-be-completed system, instead of viewing only a single aspect or rule at a time. And we all know that system integration is a much different task than that of unit development and testing.

Fig. 23 displays where these knowledge acquisition techniques fit in the spectrum of automated tools. The latter two, knowledge acquisition aid and knowledge acquisition program, will be covered in the next chapter.

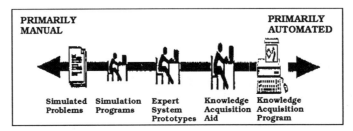

Figure 23. A Continuum of Automation for Knowledge Acquisition.
(Karen L. McGraw/Karan Harbison-Briggs, KNOWLEDGE ACQUISITION:
Principles and Guidelines, (C) 1989, p. 281. Reprinted by permission of
Prentice Hall, Inc., Englewood Cliffs, New Jersey.)

Simulations are quite familiar to most of us, as they permit experts to solve problems in an environment which is close to the real one, but not as complex. Flight simulators are common examples. The knowledge engineer can observe the actual hands-on behavior of the expert, instead of asking the latter to explicate some usually deeply-ingrained heuristics.

Prototypes are gaining rapidly in popularity. A shell of the complete system is constructed early on, and is continuously upgraded, via frequent iterations between the domain expert and knowledge engineer. Software engineers are becoming strong advocates of this approach, as the resulting system is far likelier to be a higher fidelity version of what the customer actually desires. Failing to involve the end-user in the early stages of development is inordinately expensive. This is well-known, after more than 20 years of doing precisely that. See Fig. 24. Involving the user immediately reduces the likelihood that errors in the requirements analysis will propagate to later stages. This author is a very strong advocate of this approach and feels it is a natural meeting ground for knowledge engineers and software engineers.

6. Multiple experts

The use of multiple experts is not really a knowledge acquisition technique. It is rather an increasingly popular way to assure the validity of a large KBS, by not permitting a single biased view to be reflected in the final product. It is a realization that expertise often resides with a small group of people, as well as with single individuals.

Any of the previous techniques can be used with multiple experts, either in small groups or individuals. However, there are a few additional methods which are useful to derive maximum utility from a group of domain experts. Any of these techniques can be dome manually or with the use of an automated tool.

When reviewing drafts of the McGraw & Harbison-Briggs text in early 1988, this author was pleased to see an entire chapter on multiple experts. To my knowledge, it is still the most comprehensive treatment of this important reality of today. In it, the authors list three major group techniques: brainstorming, consensus decision-making and the nominal group technique.

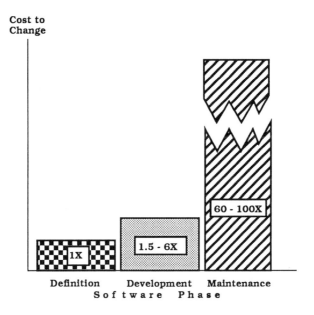

Figure 24. Rising Costs of Changing Software.

a. Brainstorming

This is the most common way for groups to interact -- the experts concentrate on generating <u>lots</u> of ideas for solving a problem without regard to the feasibility or quality. It is a method when all "knonws" are suspended and wildly-speculative ideas are the norm. No criticism is allowed, although piggy-backing is encouraged.

b. Consensus Decision-making

Here the emphasis is on determining the best solution. Advantages and disadvantages of each alternative are evaluated by each domain expert. Voting on the alternatives takes place in rounds with weaker alternatives being filtered out in the earlier voting rounds. Each person is thus assured that she/he has been heard, <u>and</u> that the final consensus is probably a good one. Only one or two options will survive the final voting round.

c. Nominal-Group Technique

This method is useful when the status of the various domain experts can be a threat, and thus would not make consensus very effective. So it is a "group" only in name, i.e., a "nominal" group. Most of the action occurs independently and perhaps anonymously.

Each member independently writes down various pros and cons to be considered. Each of these go into a pool which is rank-ordered by the group. Then every member also lists (on paper or with a computer) the various proposed solutions. These also go into a pool. Finally, the ranked pros and cons are used to investigate each proposed solution until a "best" one appears.

In summary, the array of knowledge acquisition techniques is vast and is drawn from a wide variety of fields. We have seen that some of these include lecture (from education and training), personal constructs (psychology), brainstorming (business), consensus (communications), task analysis (instructional design), protocol analysis (computer science), and prototyping (software engineering). It was obvious that we could not give even a cursory treatment to an exhaustive list of such techniques. Instead, we investigated the following six in some detail.

1. Interviews [Activity: Identify general and conscious heuristics]

    a. Unstructured Interview
    b. Structured Interview

2. Process Tracing [Activity: identify routine procedures/task]

3. Conceptual Analysis [Activity: identify major concepts/vocabulary]

    a. Taxonomies
    b. Concept Sorting
    c. Scaling
    d. Conceptual Clustering
    e. Repertory Grids

4. Task Analysis [Activity: Identify unconscious decision-making procedures and heuristics]

5. Simulations and Prototype [Activity: Identify problem solving heuristics by use of analogies]

6. Multiple experts

    a. Brainstorming
    b. Consensus Decision-making
    c. Nominal-Group Technique

Selection of Knowledge Acquisition Techniques

The original draft of this chapter had a section entitled "Trade-off Analysis of Most Common Techniques" which was to appear in this location. It was to have presented a "crib sheet" for determining pros and cons of each technique. Such a "trade-off" section is now inappropriate.

It is not so much a case of _competing_ techniques, as it is the understanding of how these various methods can _cooperate_ -- of understanding where various techniques are most useful in the development of a KBS. The well-worn figure 9 (Correlating Knowledge Type and Acquisition Technique) is very appropriate in this regard. Once it has been determined that routine procedures and tasks are to be implemented, then the knowledge engineer should use the following: structured interviews, process tracing, and simulations at various stages of development.

None of the knowledge acquisition techniques are panaceas, and they are all useful, given certain types of KBSs to be developed and the phase of construction. They all have difficulties, ranging from lack of training on the part of the knowledge engineer to the lack of appropriate hardware for various simulations. The lack of training on the part of the knowledge engineer is a major impediment to many projects. It is a major reason for the rapidly growing interest in, and use of, automated knowledge acquisition tools. This is a topic covered in detail in the next chapter. However, lacking such tools, the above difficulties are often surmounted by choosing good combinations of techniques.

Other issues in selecting knowledge acquisition techniques are common to all engineering ventures: complexity of the system, availability of developers and experts, development time, allocated finances, management backing, expected lifetime, severity of failure modes, ROI, etc.

Hence the selection or evaluation of various knowledge acquisition techniques is seen as a non-trivial task. Once again, it apparent why Bloom, back in 1955, adjudged the evaluation level of cognitive skills to be the most difficult. This is one more arena where engineering judgement comes into play!

Validation of Knowledge Based Systems

Clearly, the developer of a KBS is keenly interested in the fidelity of the knowledge which is transmitted from the source to the destination. That is, how can she easily verify that the "right" message was communicated? A more encompassing concern is the "correctness" of the entire KBS, not only the underlying knowledge acquisition methods. This topic alone deserves an entire book, as yet unwritten. So we will only touch on the subject here.

Some useful insights can be gained, once again, by looking at the world around us. How does the issue arise in "normal" knowledge acquisition? Also, what have we learned from over 20 years of software engineering experience?

First, two key definitions (paraphrased) will be helpful, one for verification, and one for validation. Although similar and often used confusingly, the concepts are different indeed:

Verification: ensures that the system is built correctly.
Validation: ensures that the correct system is built.

Verification is primarily a technical issue and is concerned with ensuring that the key concepts in one phase of the development process are faithfully mapped into the succeeding phase, e.g., from analysis to design. Validation, on the other hand, is much more user-centered -- does the delivered product actually perform what the user desired and in a manner she expected?

With these definitions in hand, let us look at applications of them in "normal" cases and then in software engineering, before investigating them for the KBS case.

1. Human verification and validation

Normally, humans acquire knowledge in a multitude of ways as detailed in Section 3. "Verification" often takes the form of grades received by the student as she moves from exam to exam within a course or from course to course, as she moves through an educational process. Enormous amounts of literature speak to this issue, probably because it is considerably easier to measure than validation. For "validation" normally takes the form of how well the person actually performs in a career after she leaves school. University accrediting agencies are very interested in the latter (SACS 1988).

2. Software engineering verification and validation

Likewise, computers acquire knowledge at various levels using a wide assortment of methods, as given in section 3. Since software engineering is currently the dominant way in which computers gain knowledge, it makes sense to investigate this area.

Software verification is one of the most active research areas today in computer science (Fetzer 1988). The proof of program correctness is an extremely popular topic <u>and</u> the source of considerable controversy, some dating back many years (Demillo, Lipton, and Perlis 1979). Wouldn't it be nice, proponents say, if the choice of proper axioms and rules of inference, made during program specification, guaranteed 100% correct code at the other end of the development phase? The majority of computing professionals, including many who have built large-scale production systems, feel this is a pipedream (Brooks 1987). Such proofs have succeeded only in very small cases of less than 1000 lines of code, and often involve proofs which are longer that the code they purport to prove! As the reader may surmise, this author is not very optimistic about the use of such formal verification methods for any realistic software products in the near or intermediate future. Generation of small secure operating system kernels would be the only exception.

Validation of a software engineering product is a much less formal issue, although of significantly more importance in the opinion of the author. After all, it really is of no consequence that a program is probably correct, unless the end user believes it actually solves a real need and in an acceptable manner. Hence, there is a growing desire to involve the end user in the software development process as early as possible. To borrow terminology from instructional design, users become involved in formative (on-going) evaluation, vs. only summative (final product) evaluation. It is <u>not</u> a pleasant experience for either party to expend enormous amounts of money and time in developing software, with no input from the user other than the initial specification, only to have the user reject the delivered product. The "correctness" of an unused program is of little interest.

This iterative and user-centered methodology is known as rapid prototyping. For a number of application domains, it is displacing the classical or "waterfall" method currently embodied in many governmental and corporate standards (DOD, 1985). See Fig. 25 and 26 for simple diagrams of these two methods.

3. Knowledge based system verification and validation.

Only recently have developers of KBSs begun to consider seriously either verification or validation of their systems. Early on, researchers at Stanford did some well-documented work by comparing the suggestions given by MYCIN for treatment of infectious bacteria with that of physicians and medical students (Shortliffe 1976). This is certainly a popular and common sense approach -- determine if output from a KBS agrees with that of human experts.

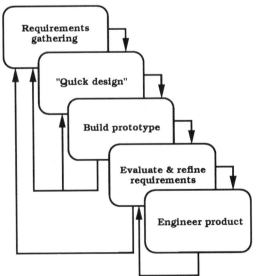

Figure 25. Rapid Prototyping Methodology.

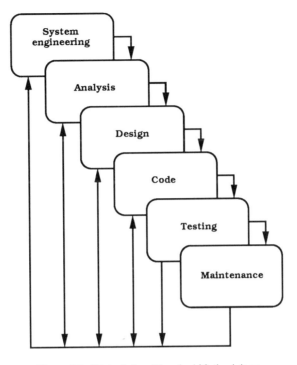

Figure 26. Waterfall or Classical Methodology.

Practitioners and researchers in KBS would be well advised to see how their software engineering colleagues address issues of verification and validation. Despite all the popularity surrounding KBSs, they are <u>still</u> software items. And, as such, managers and customers, especially those from the U.S. government, are keenly interested in seeing that software development standards are followed. The result of <u>not</u> following standards in the past is reflected in the $125 billion cost of software in the U.S. for 1990. See Fig. 27. Typically, for every software dollar spent in development, two dollars are spent to maintain and enhance the resulting product (Boehm 1987). Many people are finally beginning to realize that software development is <u>not</u> the financial driver in the software life cycle it once was.

As it turns out, KBS developers have been influential in software engineering methods. The rapid prototyping method discussed earlier is, in fact, a standard component of all KBS development -- of keeping the user very much in the loop. This is one more example of how the two fields are growing closer.

Another is the use of CASE tools to help build KBS systems. These tools help support the structure desired by managers, as well as numerous productivity and graphical aids. Several even provide automatic analysis-to-design transformation and consistency checks (McClure 1989).

The verification and validation of knowledge acquisition techniques have been shown to be part of a much larger picture. It is actually the validation of the entire KBS itself which is the issue. Analogous questions were given and answered in a traditional case for humans acquiring knowledge, e.g., schools. Likewise, the importance of the interaction between software engineering and KBS was stressed.

6. Future of Knowledge Acquisition

This section is more of a "what should be" one than "what will be." The optimism of the former stands in contrast to the "reality" of the latter. The latter -- "what will be" -- is quickly summarized by

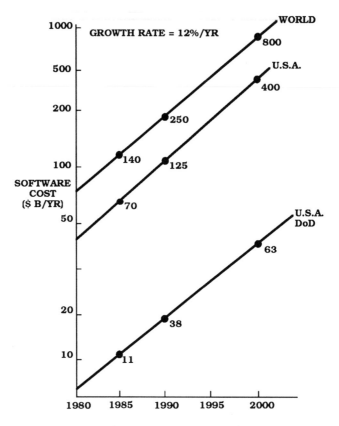

Figure 27. Rising Cost of Software.

1. More of the same of what has worked before,
2. Lots of new ideas which fail, and
3. A few new ideas which succeed.

Those ideas which will hopefully succeed -- the "what should be" -- are the following: cooperative knowledge acquisition, knowledge acquisition using a host of senses, and active knowledge acquisition.

Cooperative Knowledge Acquisition

We have touched on this earlier when discussing multiple experts in section 4. But here, it is used in the sense which NASA implied when speaking of "intelligent cooperating agents" (Merrifield 1986).

The analogy with humans is a powerful one. To be sure, difficult problems have been solved by a single person. The revered places in innovation awarded individuals such as Einstein, Edison, Goddard, DaVinci and Steinmetz are well-deserved. The persistent and intellectually creative genius will <u>always</u> have a role in helping our world be a better place in which to live. However, <u>most</u> significant large-scale problems today are ones which require teamwork. And the solutions will require some new creative people. See Fig. 28. One person <u>can</u> build a small bridge over a small creek. One person cannot build the Golden Gate bridge. One person can build a small rocket to set off in a farmer's field. It took an entire nation to land men on the moon and return them safely less than ten years after the stated intention of John F. Kennedy in 1961.

"The World that we have made as a
result of the level of thinking we
have done this far creates problems
we cannot solve at the same level of
thinking at which we created them."

- Albert Einstein

Figure 28. Solving Today's Problems.

An exciting future will be for computers to cooperate in a team environment for acquiring knowledge. Just as in human teams, the members would have a variety of complementary skills and talents. Rather that all being general purpose machines, some would excel at classification, some at hypothesis generation, some at model construction, and some at motor skills, for example.

Scotty, the space shuttle main engine (SSME) test analysis expert system mentioned earlier is a primitive example. See Fig. 29. Scotty interacted with a large multitude of other entities: humans, computers, data bases, sensors, etc. It was designed to access many sources, such as other expert systems, including ADAMEX for dynamic data analysis

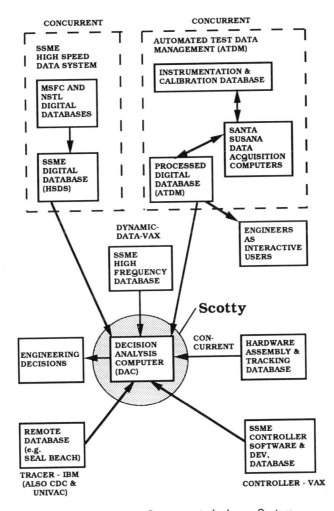

Figure 29. Scotty as Component of a Large System

(Garcia 1989). Other sources included telemetry data via T1 links, results from previous anomalies, failure mode effect analysis tables, SSME experts, and SSME engineers. Scotty was not built to have an ego about "being in charge." It was developed to solve a significant problem using a large variety of other agents, some intelligent, in a <u>cooperative</u> manner.

Acquiring Knowledge from All Senses

How courageous are those individuals who are blind, deaf, mute, or wheel-chair bound. The triumphs of a Helen Keller, Ray Charles, the aging Beethoven, or Steven Hawking are inspirational to all of us. Yet, few of us would trade places. For we normally complain when a slight cold impairs our sense of smell. We want <u>all</u> of our senses to be fully functional all of the time.

Now, suppose for a moment, that you are being requested to acquire knowledge about the entire world around you. Suppose further that you are locked inside a digital computer. What ways are available for you to sense and affect the world? For decades, the answer has been "any that can be encoded as (usually) digital signals." Most of the time, this has meant numbers and characters. Newell's physical symbol hypothesis expands this considerably (Newell 1976).

More recently, advances in CCD imaging have permitted pictures to be transmitted to and from you, still locked in the computer. The encoding/decoding of signals could permit you to detect and affect motion changes, as in robotics. Also recently voice recognition as well as the easier voice synthesis has become a commercial reality, e.g., Bell's Freedom Phone.

Let's see where you currently are in your computer box with respect to sensors and affecters. See Fig. 30:

SIGHT: read numbers, letters

see simple pictures

detect motion

HEARING: recognize voices in a quiet environment

TOUCH: can assemble some simple devices

SMELL: not yet, except maybe in some laboratories

TASTE: not yet, except maybe in some laboratories

MOVEMENT: can navigate around relatively simple
environment, usually flat and with few
obstacles (assuming wheels or primitive legs)

SPEAKING: can make great music via synthesizer!

FEELINGS: only in science fiction

Figure 30. Current State-of-the-Art for a Sensory Computer.

The author has no illusion of being an expert in all of these fields, e.g., robotics, computer vision, speech, natural language recognition, etc. However, the summary is probably not far astray. How does it feel in your box now? Moreover, let's make the situation even less tenable. Instead of being an adult, suppose you were virtually a newborn, in terms of your mental faculties, and yet still locked inside the computer. Looks pretty bleak, doesn't it, when you are still charged with "acquiring knowledge of the world?"

Why is it that we expect such a terribly handicapped entity to acquire much knowledge about our dynamic world? Would we expect much of a similarly impaired human being, even if they did have gigabytes of memory and could process billions of instructions per second? Clearly, this anthropomorphism can be extended too far, but the point should be obvious. The input and output channels of computers will most likely be altered dramatically before we can realistically anticipate much knowledge acquisition to occur.

Active Knowledge Acquisition

When was the last time that a professional colleague or student asked you a probing question, forcing you to examine and explain your reasoning clearly and succinctly? Later on, you saw that same colleague or student incorporate your answer with her own experiences (heuristics and deep models), becoming a better problem solver as a result of the interchange with you? Now, when was the last time a computer asked you a similar question? Did you notice a consequent behavior change on the part of the machine?

Let's take another aspect, that of curiosity. Don't you often try something new, just to see what happens? Oh, certainly, engineers are normally conservative, and usually proceed cautiously. However, this author is willing to wager that your curiosity is more responsible for your professional success than the logical part of you cares to admit. In fact, this same curiosity may well have resulted in your choice of this very profession. It has been said that our curiosity is actually a stronger drive than our sexual one. This very curiosity allows us to uncover relationships (analysis) and generate ideas (synthesis) in ways not possible purely by logical and rational processes.

When was the last time you saw a truly "curious" computer, in the sense used above? Because of its handicapped sensory motor skills, has one ever asked you to perform some untried experiment? Did it then ask you to tell it what happened? Note that this is not the same as aimless trial-and-error. Humans are not normally aimless -- they have some purpose in mind, even though it probably cannot be articulated. Only a minute amount of research effort has been expended in this type of "discovery learning" by computer. Some success had been reported in discovering "new" theorems and unique 3-D VLSI circuits (Lenat 1976).

Summary

Cooperation, use of all senses and affecters, and curiosity all play extremely vital roles in the lives of human beings, particularly as they acquire knowledge to solve significant problems. And there is a very strong interaction among all three.

How often does exploration in one sense impact curiosity along another dimension? "So this is how X looks. I wonder how it smells." "If Y can move so fast, I wonder what delta energy change occurs?" "Since vanilla flavoring smells so wonderful, I bet it tastes great -- let's try it!"

Teamwork also is an interacting agent. "Since this team of five people does such excellent work in developing software, I bet that a group of fifty people can do ten times better." Wrong! See the classic Myths of Software Engineering (Brooks 1975)

These three then: cooperation, senses and curiosity, are this author's nominations for some of the survivors among the many ideas which will be espoused in the near future for knowledge acquisition.

7. Summary

In this chapter, we have attempted to give an overview of several aspects of knowledge acquisition. This overview consisted of exploring the why, what, how, and where of knowledge acquisition. We have tried to articulate some basic principles and techniques of knowledge acquisition which have worked well in the past. In particular, we explored those which have involved human beings as the intermediary. This intermediary, usually called a "knowledge engineer," is the agent for the transmission of some type of knowledge from a knowledge source to a destination, normally with considerable "noise" added during the transmission.

Why are we interested in acquiring knowledge? Because we wish to do a better job of solving problems -- of filling gaps between what is currently and what could be in the future by changing our heuristics over time. We wish our functional needs to be satisfied effectively and efficiently.

The "what" of knowledge acquisition was investigated from the context of various types of knowledge which can be acquired. We discussed six categories: knowledge, comprehension, application, analysis, synthesis, and evaluation. Then an earlier definition for knowledge acquisition was extended: "the translation and transformation of problem solving expertise from a knowledge source (e.g., human expert, documents) to a human or computer program destination." Common examples were drawn from the likely fields of education and work (training and performance).

The "hows" of knowledge acquisition were next undertaken, both from the viewpoints of humans as acquirers of knowledge, as well as computers. For each, the relevant methods were divided into past, present, and some future ones. Time itself was seen to be critical over the lifetime of a learning entity, as progress was made from being a student or novice to becoming a knowledgeable and hopefully wise expert.

The "practical" part of the chapter was in a "heuristic cookbook," drawing upon the key work, *Knowledge Acquisition: Principles and Guidelines*. In this part, the major techniques which are currently used by knowledge engineers were summarized and compared. Validation Issues were also covered.

Future issues in knowledge acquisition were discussed, including the role of cooperative systems (cooperative people, cooperative computers). Other issues addressed the difficulty of a machine having all the senses which a human being uses in her knowledge acquisition, as well as the requirement to have the computer be curious -- to ask questions freely.

8. So What?

Roger Kaufman, Director of the Center for Needs Assessment and Planning at Florida State University, once visited Texas Instruments Incorporated at the request of the author, an employee there at the time in the early 1980s. During his consulting visit, Professor Kaufman made many excellent suggestions based on his prolific work (Kaufman 1982). Two of the ones relevant here were his stressing the importance of always asking "why?" and "so what?" In the context of knowledge acquisition, the "why" part has been addressed to a considerable extent. Now for the "so what" component.

In some respects, the two queries are similar. The "so what" can be paraphrased as "what difference does all of this make if the idea does catch on the succeed?" What impact will occur if (when) we are able to help machines be successful in acquiring knowledge?

Will these machines then replace people (Garis 1989)? Without a long digression, the answer is "yes" and "no." Did tractors replace people? Yes, some field hands were replaced. No, some field hands were trained to drive tractors. No, field hands and others were taught how to design, build, sell, and repair tractors. Did electronic switching systems replace people? Analogous answers can be given. Some are replaced and some of these retrained. Many more new types of positions are also generated.

Will computers which acquire knowledge become Frankensteins or worse? They certainly have the potential. The decision is up to us, the designers and users. The computer itself is amoral. Hammers can be used to hit someone, or to build towering cathedrals to the glory of God. Computers most certainly are not created to destroy!

Will computers with knowledge beyond that we take for granted today suddenly become omniscient and incapable of failure? Hardly! In fact, from our long history as humans exercising heuristics, we <u>know</u> that any such computers <u>will fail</u>! In the future, the impact of such failure will have wide-ranging consequences, many negative. In the United States, we have only to look at the aftermath of the failure of one of our most marvelous machines, the space shuttle Challenger in January, 1986. The learning which transpired as a result of this tragic failure now makes space travel much safer, although clearly not 100% safe.

Will imparting more knowledge to computers have unanticipated consequences? Yes!

9. Conclusions

Knowledge begets knowledge. However, Richard Hamming once said at an ACM national conference in 1969: "In other fields, people build on one another's shoulders -- in computing we build on one another's toes!" We are trying to move beyond this state. This chapter is an attempt to build on the shoulders of others: engineering, instructional design, education, software development, and others. We hope the result has been an increase in knowledge for you. Even more fervently, we hope that it will be used with <u>wisdom</u>. Perhaps the next work will be on wisdom acquisition...

Acknowledgements

Thanks to Dr. Tomasz Arciszewski for prompting me to write such a chapter. Dr. Karen McGraw gave me a great honor by letting me review her seminal work, as well as write the preface -- I shall never forget the pleasure of doing this. Dr. Roger Kaufman has been a mentor for over ten years, and I hope he finds some of my adaptations of his important ideas explained clearly herein.

Most of all, Jan Modesitt is the primary person who has given me support and encouragement for me to be the best I can be. Thank you...

Appendix-References

Bloom, B., (1956). *Taxonomy of Educational Objectives. Handbook I, Cognitive Domain*, David McKay, New York.

Boehm, B. (1987). "Industrial Software Metrics Top 10 List," *Software*, IEEE, September, pp. 84-85.

Boose, J., (1989). "A Survey of Knowledge Acquisition Techniques and Tools." *Knowledge Acquisition,* Vol. 1, pp. 3-37.

Broad, W., (1990). "Wanted on the Space Station: Better Suits, Robots, and Parts," *NY Times*, March 27, pp. B5-B8.

Brooks, B., (1987). "No Silver Bullet: Essence and Accidents of Software Engineering," *Computer*, IEEE, April, pp. 10-19.

Brooks, F., (1975). *Myths of Software Engineering*.

DeMillo, R., Liption R., and Perlis A., (1979). "Social Processes and Proofs of Theorems and Programs," *Communications of the ACM*, Vol. 22, No. 5, pp. 271-280.

DOD, (1985). *Defense System Software Development*, DOD-STD-2167A.

Dreyfus, H. and Dreyfus S., (1986). *Mind over Machine: The Power of Human Intuition and Expertise in the Era of the Computer*; Free Press, New York.

Ellman, T., (1989). "Explanation-Based Learning: A Survey of Programs and Perspectives," *Computing Surveys*, ACM, Vol. 21, No 2, pp. 163-221.

Feigenbaum, E. and Barr A. (1989). *The Handbook of Artificial Intelligence*, William Kaufman,Inc, Los Altos, CA.

Fetzer, J., (1989). "Program Verification: The Very Idea," *Communications of the ACM*, Vol 31, No. 9, pp. 1048-1063.

Ford G. and Gibbs N., (1989). "A Master of Software Engineering Curriculum," *Computer*, IEEE, September, pp. 63-71.

Garcia, R. (1989). "ADAM-EX: The Use of Dynamic Data and Inductive Methods to Automate the Analysis of Reusable Propulsion Components," *Proc. of Workshop on Inductive Programming*,IJCAI, Detroit, MI.

Garis, H. (1989). "What if AI Succeeds?: The Rise of the 21st Century Artilect," *AI Magazine*, Vol. 10, No. 2.

Harmon, P. and King D., (1985). *Expert Systems*, Wiley.

Hayes-Roth, F., Waterman D., and Lenat K., (1983). *Building Expert Systems*, McGraw-Hill.

Kaufman, R., (1982). *Identifying and Solving Problems.* Third Edition, University Associates, San Diego, CA.

Kelly, G., (1985) *The Psychology of Personal Constructs*, Norton.

Koczkodaj, W., (ed.) (1991). *International Journal of Software Engineering and Knowledge Engineering,* World Scientific, London.

Koen, B., (1985). *Definition of the Engineering Method,* American Society for Engineering Education, Washington, D.C.

Koen, B., (1986). "The Engineering Method and the State-of-the-Art," *Engineering Education*, April, pp. 570-674.

Lenat, D., (1976). *An Artificial Intelligence Approach to Discovery in Mathematics as Heuristic Search*, Ph.D. Thesis, Stanford University.

Mandl H. and Lesgold A. (eds.) (1989). *Learning Issues for Intelligent Tutoring Systems*, Springer-Verlag.

McClure, C., (1989). "The CASE Experience," *BYTE*, April, pp. 235-246.

McGraw, K. and Harbison-Briggs K., (1989). *Knowledge Acquisition: Principles and Guidelines*, Prentice-Hall.

Merrifield, J. (1986). Ames Readies Initial AI Demonstration for Space Station," *Aviation Week & Space Technology*, May 26, pp. 129-131.

Michalski, Carbonell J.G., and Mitchell T., (eds.) *Machine Learning*, Tioga Publishing.

Michie, D., (1987). "How Artificial Intelligence Fits In," *The Turing Institute Reports*. Glasgow, Scotland.

Modesitt, K., (1990a). "Computer-based Learning, Expert Systems, and Software Engineering: Advanced Hybrid Tools for Engineering Education Now and in 2001," *Frontiers in Education Conference*, American Society for Engineering Education, Vienna.

Modesitt, K., (1990b). "Inductive Knowledge Acquisition Experience with Commercial Tools for Space Shuttle Main Engine Testing," *Proc. of Fifth Conference on Artificial Intelligence for Space Application*, NASA/University of Huntsville, AL.

Modesitt, K, (1989). "The Integration of Automated Knowledge Acquisition with Computer-Aided Software Engineering for Space Shuttle Expert Systems," Workshop on Knowledge Acquisition, *Proc. of Intl. J. Conference on Artificial Intelligence*, Detroit, MI.

Modesitt, K., (1988).   "Experience with Commercial Tools Involving Induction on Large Databases for Space Shuttle Main Engine Testing," *Invited talk for Fourth International Expert Systems Conference*, London, England, June, pp. 219-229.

Modesitt, K., (1987). "Experts: Human and Otherwise," *Proc. of Third International Expert Systems Conference*, London, England, pp. 333-342.

Modesitt, K., (1983). "Computer Based Learning: Important Problems, Creative People, and Powerful Affordable Tools," *Journal of Computer-Based Instruction*, Vol. 9, May, pp. 26-33.

Myers, W., (1989). "Allow Plenty of Time for Large-Scale Software," *Software*, IEEE, July, pp. 92-99.

NASA, (1985). "Advancing Automation and Robotics Technology for the Space Station and for the U.S. Economy," *NASA Technical Memorandum 87566*, March.

Newell, A. and Simon H., (1972). *Human Problem Solving,* Prentice-Hall.

Newel, A. (1976). "Computer Science as Empirical Inquiry: Symbols and Search," *Communications of the ACM*, Vol. 19, pp. 113-126.

NTU, (1990). "NTU Celebrating Five Years," *NTU UPLINK*, National Technological University, Vol. 5., No. 4, April.

NTU, (1989). "New Software Engineering Series Complements Corporate Training," *NTU Uplink*, National Technological University, August.

Papert, S., (1980). *Mind Storms: Children, Computers and Powerful Ideas*, Basic Books, New York.

Patterson, D., (1990). *Introduction to Artificial Intelligence and Expert Systems*, Prentice-Hall.

Piaget, J., (1950). *The Psychology of Intelligence*, Harcourt, New York.

Polya, G. (1973). *How to solve it: A New Aspect of Mathematical Method*, 2nd edition, Princeton University Press.

Rich, C. and Waters R. (eds.) (1986). *Readings in Artificial Intelligence and Software Engineering*, Morgan Kaufman, Los Altos, CA.

SACS, (1988). *Criteria for Accreditation: Commission on Colleges*, Southern Association of Colleges and Schools.

Sanger, S., (1990). "Japan Launches Rocket to the Moon," *NY Times*, January 25, p. I1.

Schmitt, H., (1986). "A Millennium Project: Mars 2000," *The Planetary Report*, The Planetary Society, July/August.

Shortliffe, E. (1976). "Whether Software Engineering Needs to be Artificially Intelligent," *Transactions on Software Engineering*, IEEE, Vol SE-12, No. 7, July.

Taguchi, G. and Phadke M.S., (1984). "Quality Engineering Through Design Optimization," *Proceedings of the Globecom Meeting*, IEEE Communication Society, pp. 1106-1113.

Warman, D. and Modesitt K., (1988). "A Student's View: Learning in an Introductory Expert System Course, *Expert Systems: Intl. J. of Knowledge Engineering*, pp. 30-39.

Waterman, D., (1986). *A Guide to Expert Systems*, Addison-Wesley.

Welch, D., (1989). "*Star Channels*," KET, The Kentucky Network.

CHAPTER 3

Machine Learning in Knowledge Acquisition

Tomasz Arciszewski
Wojciech Ziarko

1. Introduction

This chapter provides a brief introduction to machine learning in knowledge acquisition. Section 1 explains why machine learning is so attractive to knowledge engineers. Section 2 discusses basic concepts of machine learning and learning systems. Section 3 covers various taxonomies of machine learning, including three classical taxonomies and a recently-proposed one based on classification by coverings using eight binary logical attributes. Section 4 discusses several major approaches to learning from examples, which is particularly important in engineering. Section 5 provides a comparison of three inductive systems, conducted using the same collection of examples from the area of civil engineering. Section 6 briefly describes future developments in machine learning and knowledge acquisition.

As discussed in the preceding chapter, learning is closely associated with both human and machine intelligence. Therefore no intelligent system can exist without the ability to learn. Dreams about learning computers are as old as research on computing and artificial intelligence. In 1950, Allen Turing, one of the fathers of computing, wrote his now famous paper, "Computing Machinery and Intelligence" (Turing 1950). In this paper he discussed different aspects of intelligence as related to computers, and mentioned for the first time the concept of learning machines. This concept captured the imagination of researchers, and led to more than forty years of work on machine learning, as described in Chapter 2. Only recently, machine learning research resulted in the development of a class of commercial and experimental learning systems which are fairly user-friendly and can produce useful results. These systems can also be applied in civil engineering for knowledge acquisition. Therefore, an understanding of the underlying principles of these systems, and their potential applications and limitations, is of significant importance to everyone interested in knowledge acquisition. This chapter presents a brief description of these systems. The discussion includes examples of the application of several learning systems, to help the reader understand the learning principles utilized and to determine her/his potential preferences.

The use of machine learning in knowledge acquisition is as yet neither simple nor routine. A knowledge engineer should be properly motivated and should know both the advantages and disadvantages of machine learning compared to human knowledge acquisition. Knowledge acquisition is defined in Chapter 2 as:

"... the translation and transformation of problem solving expertise from a knowledge source to a human or computer program destination."

There is, however, another definition (Michalski et al. 1983) which clearly reveals the importance of learning in knowledge acquisition:

"Knowledge acquisition is learning new symbolic information coupled with the ability to apply that information in an effective manner."

This definition covers both human and machine learning. We are interested here in machine learning, and for the purposes of this chapter we will assume that the knowledge source is given data, which is transformed by a learning system into knowledge. This knowledge can be used in a knowledge-based system. The development of such a system can be considered as a two-stage process involving knowledge acquisition and the implementation of the acquired knowledge. In this chapter we will concentrate on knowledge acquisition.

In the area of knowledge acquisition, learning systems should attempt to achieve one or more of the following goals (Forsyth and Rada 1986):

1. Provide more accurate decision rules
2. Cover a wider range of problems
3. Obtain answers more economically than humans
4 Simplify codified knowledge

The first two goals are the most important. Simplification of the available knowledge is understood as its rearrangement to become more intelligible to humans. Even if the system's performance has not been improved, the simplification of knowledge is generally useful.

Many expert system development shells are presently available. All these shells have tools for knowledge implementation, but most of them lack tools to support knowledge acquisition. Therefore the use of these shells requires using traditional, or manual, knowledge acquisition techniques. These techniques are not simple, and they usually require a time-consuming process, as described in Chapter 2. This process of knowledge acquisition is a bottleneck in the development of knowledge-based systems (Barr and Feigenbaum 1981). Therefore any progress in the area of knowledge acquisition will have a significant practical impact.

Most of the available expert system development shells require providing knowledge in the form of decision rules based on a number of attributes. Unfortunately, this form of knowledge representation is not natural for human experts. It is estimated that only one percent of human knowledge is in the form of explicit decision rules, and usually a human does not act according to a well-defined set of rules or algorithms. However, human experts, particularly those with many years of experience, are capable of drawing correct conclusions by recognizing characteristic patterns or dependencies in their domains. However, they may not be able to describe the reasoning procedures they used to arrive at a conclusion, nor to identify the decision rules they were subconsciously using. A typical example is bridge maintenance, where a bridge expert recognizes the condition of a bridge rather than deduces it from a set of predefined rules. Since human experts are not able to identify their decision rules in an explicit form, rules acquired from experts usually lead to an imperfect system of rules in terms of completeness and correctness. The performance of a knowledge-based system using these rules will always be below the performance of the human expert who provided the rules. This deficiency of systems built using manual knowledge acquisition techniques becomes particularly important in complex domains, where the use of knowledge-based systems is especially desired and justified. It is also known that human experts usually eagerly provide examples of their correct and incorrect decisions, even if they are not willing or able to explain the reasoning which led to these decisions. All such examples can be used as a source of knowledge, provided that there is a tool for converting these examples into a system of decision rules.

Machine learning has significant advantages over human learning. The two types of learning can be compared by considering learning from examples, or inductive learning, which is the best-understood form of learning. Humans are very good at deduction, at using available general knowledge for dealing with individual problems. However, we have very limited inductive abilities. By induction, we mean the development of general knowledge, or decision rules, from examples. Humans can handle only a very limited number of examples and attributes of a problem at the same time, because of the limitations of the human working memory. The average human can handle only seven attributes and seven examples at a time, while a properly programmed computer can deal with large numbers of both attributes and examples, limited only by the available working memory.

A knowledge-based system in a complex engineering domain may have hundreds of decision rules. Any manual changes or modifications of such a knowledge-base that maintain consistency are time-consuming and can be compared to modifications of a large algorithmic program. Clearly, errors are inevitable in this process, making manual system updates even more difficult. The use of inductive learning radically improves the situation. It requires the development of several examples illustrating new conditions and the repetition of the induction using old and new examples together. The inductive system produces a new system of decision rules which can be used immediately, after its validation, in the knowledge base.

The last, but not least, major problem with knowledge-based systems is their inability to acquire knowledge in the absence of human experts. This problem occurs when dealing with incompletely understood processes, for example in the area of waste treatment or earthquake engineering. The required knowledge cannot be obtained entirely from human expert, and only the use of machine learning can help produce the necessary decision rules.

The existing limitations on manual knowledge acquisition, briefly discussed here, are the main reason why the search for novel methods of knowledge acquisition is so important. Progress in machine learning may change the present practice of knowledge acquisition, and may lead to the development of a new generation of methods based on the use of different learning tools.

2. Basic Concepts

There are many definitions of learning, but we choose the one proposed by Simon (1983):

"Learning denotes changes in the system that are adaptive in the sense that they enable the system to do the same task or tasks drawn from the same population more efficiently and more effectively next time."

This definition relates learning to system adaptability, which is a common feature of all learning systems. A learning system capable of self-tuning is also expected to improve its performance, and these improvements should be measurable. This fact is clearly stated by Forsyth: "If we cannot evaluate a system's performance, we cannot say whether it has learned anything" (Forsyth and Rada 1986). Unfortunately, the process of monitoring and evaluating a learning system's performance is very complex and will not be discussed here, although some initial strategies to deal with this process have already been developed (Arciszewski and Mustafa 1989, Mustafa and Arciszewski 1989).

A learning system for knowledge acquisition can be considered as a black box. It transforms a given input into output. In all known learning systems this transformation is based on the common paradigm:

Generate and Test a Hypothesis

This is also the paradigm of all scientific discovery (Popper 1959). A hypothesis is a concept, or a decision rule. The input is understood as given data, usually a collection of examples. The output is usually in the form of decision rules, with the exception of systems using explanation based-learning, as discussed in the next section.

The learning paradigm can be explained by considering a simple model (Fig. 1) of a learning system for knowledge acquisition acting in the incremental mode. This system transforms given data into knowledge. The input data and the output knowledge may be in different forms.

The input data is divided into training and test data. The training data is used for learning, while the test data is intended for testing and evaluation of the knowledge produced as the result of learning. There are two basic components of the system: The Knowledge Generator and Knowledge Tester. The knowledge Generator receives the training data and conducts learning: it extracts the initial knowledge from the training data. This knowledge is then sent to the Knowledge Tester, which tests and evaluates it using the test data. The results of these tests are returned as evaluated knowledge to the Knowledge Generator. Learning is repeated, if necessary, or the initial knowledge becomes final knowledge and is sent as output to the system's user. The feedback between Knowledge Generator and Knowledge Tester is important: it causes the system to operate in a closed loop until the knowledge produced is sufficiently accurate (if possible) and can be given to the user. There are learning systems, e.g. DataQuest (Reduct Systems, Inc. 1990), discussed in Section 4, which conduct learning in one shot and produce results without internal testing and refinement. In the case of such systems, the input data remains undivided, the Knowledge Tester must be assumed inactive and the Initial Knowledge is presented to the user as output.

Machine learning involves three basic types of inference processes: deduction, induction, and abduction. Inference can be defined as a process of logical reasoning which produces new knowledge from existing one (Firebaugh 1988).

Deduction is an inference process using general knowledge to deal with particular cases. The inference is usually based on the use of the rule of detachment, called modus ponens. (If for any two schemata A and B the schema A is true and the implication A--->B is true, then the schema B is true.) Therefore deduction is truth-preserving. It can be used in knowledge-rich learning, when a large body of knowledge is available and can be used to guide the learning process. Deduction is also utilized in explanation-based learning, described in the next section.

Induction is the inference process of formulating general knowledge by dealing with particular cases. Induction is logically invalid and is falsity-preserving, but it can produce accurate results when dealing with a sufficiently large number of particular cases.

Abduction is an inference process based on reversing time implications. It is falsity-preserving. An abductive conclusion is always consistent with available information, but may be false (Kondratoff 1990). To illustrate abductive reasong a simple implication can be considered:

If a given structure is a truss, this implies that the structure has straight members.

The abductive conclusion will be: if a given structure has straight members, then it is a truss.

There are different forms of input data and knowledge representation, discussed in the next section under the taxonomy of machine learning systems based on types of knowledge. However, in engineering applications the most convenient method is usually to use attributes for both the input data and the final knowledge. These attributes may be of a nominal or numerical character. Nominal attributes are variables whose values are of symbolic character, i.e., for an attribute = material, the values may be steel, wood, concrete, etc. Numerical attributes obviously have numerical values, and do not require any explanation.

3. Taxonomy of Machine Learning

There are three classical taxonomies of machine learning as proposed by Michalski et al. (1983). These taxonomies are based on considering three different features of learning systems, namely:

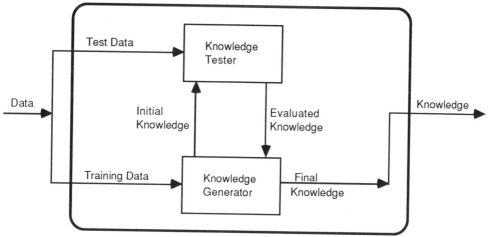

Figure 1. A Learning System for Knowledge Acquisition

1. Learning strategies
2. Representation of produced knowledge
3. Application area

LEARNING STRATEGIES

The taxonomy based on learning strategy classifies learning systems according to the amount of inference the system performs on the data provided. The concept of the amount of inference can be explained by considering calculations based on algorithmic programming and unsupervised learning. These two types of computer operations represent the ends of a spectrum as far as the amount of inference is considered. In the first case a computer is programmed by a human programmer and its operations are entirely programmed and based on the reasoning of the programmer. The use of an algorithmic program transforms given data into knowledge, but this transformation is conducted without any inference by the computer. In the second case of the unsupervised learning the system is provided with raw data from experiments, and transforms this data into new theories. The discovery of these new theories, based on the analysis of the raw data, obviously requires a very significant amount of inference which cannot be entirely predicted by a human programmer.

There are eight basic types of learning, arranged according to the growing amount of inference:

1. Rote learning
2. Learning from instruction
3. Learning by analogy
4. Case-based learning
5. Learning from examples
6. Neural networks-parametric learning
7. Explanation-based learning
8. Learning from observation and discovery

Rote learning is the simplest form of learning. It is like memorization of the experience for reuse in future situations. It does not require any inference or transformation of knowledge. The process of learning is reduced to storing facts without any generalization. The learning system has a data base and a pattern matching and information retrieval mechanism to conduct learning, i.e., to store the given decision rules. The best known experimental implementation of rote learning is in Samuel's program for playing checkers (Samuel 1959).

Learning from instruction requires only a limited amount of inference. The input data is in the form of instruction or advice. An advice statement is a heuristic containing general knowledge which must be transformed by the system itself into an executable procedure. The learning system has a data base and a system of operators to conduct learning, i.e., to convert a given general decision rule into a procedure, into an executable sequence of decision rules. The concept of learning from instruction was used in FOO, short for "First Operational Operationalizer" (Mostow 1983).

Learning from analogy requires more inference than the two types of learning discussed above. The input data contains knowledge related to the behavior of an abstract or real object, and the learning system transforms it into knowledge related to a similar object in terms of its behavior or structure. Reasoning by analogy is a powerful form of human inference, and this is one of reasons why machine learning by analogy is considered particularly promising and has been used in a number of computer programs (Burstein 1986, Kedar-Cabelli 1986).

Case-based learning is a learning process conducted using a database containing classified examples, past cases. This database is used directly to classify new cases without the prior extraction of the classification rules from the past cases, as it is in the other types of inductive learning. A new case is compared to cases stored in the database to find a past case which is most similar, or is the closest one in terms of a predefined similarity measure, to the case considered. The classification of a new case is thus based on the direct analysis of the past cases.

Learning from examples, from experience, or inductive learning, requires a significant amount of inference when compared with other types of learning. The input data is in the form of a collection of past classification decisions describing the past behavior of a given abstract or real system. Examples are in terms of nominal and/or numerical attributes and their values. The learning system transforms the given collection of examples into classification rules, or decision rules. These decision rules are generalizations based on the examples. Inductive learning is particularly useful for the purposes of knowledge acquisition in engineering, and therefore several inductive learning strategies and their implementations are discussed in the next section.

In spite of the exciting name "neural nets," the artificial neutral nets do not have much in common with real neural nets functioning in living organisms. Artificial neural nets are computational procedures composed of simple elementary functions such as summation and multiplication, aimed at modeling more complicated functions. The function F which a neural net attempts to model must be given in the form of a sequence of $(x_i, F(x_i))$ function input and output pairs. The sequence of such pairs, which can be represented as an attribute-value table, is used to control the neural net generation process by using a network training algorithm. In this process the training algorithm goes through the training set of function data a number of times while gradually improving the computational procedure to represent the function F. Depending on the types of function F of the training algorithm and the parameter setting, net generation may take a very long time, ranging from minutes to months on a standard computer, or may not be successful at all. To overcome this inherent inefficiency of network training algorithms, a number of hardware manufacturers have come up with expensive parallel processors to speed up the generation process and make the technique applicable to real-life problems.

Research in the area of neural networks was initiated with the introduction of the perceptron by Rossenblatt (1987). The perceptron can be seen as a classification device which, given a set of preclassified numeric training input patterns, is able to synthesize a corresponding linear discriminant function. This function has the same value for all training examples belonging to the same class. Any two patterns from two different classes are assigned different values. In this case the patterns can be classified simply by computing the value of the discriminant function. The perceptron, however, has one fundamental limitation related to the fact that the disriminant function produced by the perceptron is a combination of linear functions with a simple threshold function.

Explanation-based learning is a specific type of learning where no entirely new knowledge is acquired. Instead, available knowledge is refined or generalized. The process of learning is based totally on inference, using a complete, or nearly complete, knowledge base for the learning domain. The knowledge base contains a consistent system of decision rules, or theorems, and facts. The learning system is given an instance of a new decision rule, or of a potential new theorem, expressed in terms of attributes and their values. The system verifies the consistency of this rule with the rules already in the knowledge base. If a given example of an unknown decision rule satisfies the consistency requirement, the decision rule, or new theorem, is added to the knowledge base. The learning process is a mechanical theorem-proving process which may use decision trees, called explanation tress, for proving decision rules (Adeli and Yeah 1990, Harandi and Lanye 1990, Kodratoff 1990).

Learning from observation, also called discovery, or unsupervised learning, is a type of inductive learning requiring a large amount of inference. The input is in the form of a very large number of examples, obtained from continuous monitoring of an object. The learning system transforms the input into decision rules, or even attempts to generalize the input into a behavioral theory. The best-known computer program using learning by discovery is EURISCO. This program has been used in the area of biology, where it discovered a number of useful heuristics related to the problem of DNA analysis (Lenat, Michalski 1983).

KNOWLEDGE REPRESENTATION

In the taxonomy based on the type of knowledge acquired, learning systems are divided into ten basic types (Michalski 1983). This classification is based on the form of representation of the knowledge produced by the system. The identified types and their brief descriptions are as follows:

1. Parameters in algebraic expressions.

The learning system produces knowledge in the form of values of parameters in algebraic expressions of a fixed mathematical model developed for pattern recognition. This type of knowledge representation is used in many systems based on parameter adjustment, including simple systems using discriminant function or signature tables, and very complex neural networks.

2. Decision tree.

This is a popular form of knowledge representation in inductive systems. Individual nodes in the tree represent attributes and the edges correspond to values of these attributes. An example of a decision tree produced by the inductive system Super-Expert (Intelligent Terminals 1986) is given in Section 4.

3. Formal grammars.

A learning system for the recognition of a language produces knowledge in the form of formal grammars induced from the sequences of sentences (expressions) in a given language.

4. Production (decision) rules.

This is the most popular form of knowledge representation in inductive systems. The results of the learning are in the form of implications, or condition-conclusion (action) pairs expressed in nominal and numerical attributes. Examples of such decision rules from the system BEAGLE (Warm Boot 1987) are given in Section 4. (Winston 1977).

5. Formal logic-based expressions.

The knowledge produced by a system is in the form of formal logical expressions (formulas) built using propositions, predicates, and attributes (Nilsson 1971).

6. Graphs and networks.

The learning system produces knowledge in the form of graphs or networks. Networks are equivalent to formal logical expressions but are more convenient in engineering applications.

7. Frames and schemata.

A frame is a date structure that includes declarative and procedural information in predefined internal relations (Barr and Feigenbaum 1981, Minsky 1975).

A schema is a logical expression built using formulas and logical connectives. It is more general than a single logical formula.

8. Procedural representations.

A learning system for automatic programing produces knowledge in the form of a computer code (Woods 1968).

9. Taxonomies.

Learning from observation may produce a hierarchical classification of the objects considered, i.e., a taxonomy.

10. Multiple representation.

A learning system may produce knowledge in several different forms, for example production rules and taxonomy.

APPLICATION AREA

A taxonomy based on the application domain is obvious. Learning systems are classified according to the application area in which they are used, for example, general (no specific domain), civil engineering, cognitive modeling, game playing, robotics, mathematics, etc.

GENERAL CLASSIFICATION OF LEARNING SYSTEMS

The preceding section discussed a general model of a learning system for knowledge acquisition. This model can be used to develop a classification of learning systems based on eight nominal binary attributes that describe the system and its operations. These attributes and their values are given in Table 1.

When a given attribute is considered, its two identified values represent the ends of its variation spectrum. Quite often, for a real system, the value of a given variable falls somewhere between these two extremes, and selecting only one of the two assumed values may oversimplify the classification.

The classification of a given system may be assisted by the use of a checklist (Forsyth 1989) containing two-way questions pertaining to the individual attributes given in Table 1. These questions are as follows:

1. Does the system apply to a variety of application areas or is it specialized for one field?

Table 1. Taxonomy of Learning Systems: Logical
Attributes and their Values

	Attributes		Values	
Domain	A1	Generality	General Purpose	Specific
Learning Process	A2	Type of Learning	Incremental	One-Shot
	A3	Degree of hierarchy	Subsumption-based	Non-hierarchical
	A4	Determinicity	Deterministic	Non-deterministic
Evaluation	A5	Evaluation	Logical	Quantitative
Representation	A6	Examples	Unary Features	Logical Predicates
	A7	Transparency	Humanly Intelligible	Black Box
	A8	Language	Fixed	Extensible

2. Does the system maintain a best-so-far rule of description which is amended as new examples arrive one by one, or does it look at the entire training set before forming its rule(s)?

3. Does the induction process rely on the inheritance lattice that orders concepts from general to specific, or does it work without reference to the generality of the descriptions it generates?

4. Does the system always give the same results from the same data, or is there a random element in its rule generation?

5. Does the system classify trials only as successes or failures, or does it measure the distance from the correct answers according to a metric?

6. Are the examples presented to the system as feature vectors, or can the system accept examples which have internal structure described by multi-term predicates?

7. Is the rule language readable by people, or it is in an opaque internal code?

8. Is the system restricted to a description language given by its designer, or can it extend its own vocabulary by defining new concepts and functions?

Usually, the description of commercial learning systems is incomplete, and it is difficult to classify them properly. However, the identified attributes and the checklist are a convenient aid for knowledge engineers in the classification of new learning systems and their comparison, at least partially, with existing ones.

4. Selected Inductive Knowledge Acquisition Tools

This section describes four commercially available knowledge acquisition tools employing learning from examples. All these tools discussed in this section were initially developed for experimental purposes. However, they proved to be useful in practical applications in knowledge acquisition, and commercial versions were prepared later.

BEAGLE

BEAGLE is a software package developed by Warm Boot Ltd. of Nottingham, United Kingdom. The learning process is based on genetic algorithms, proposed by Holland (1975) and described in engineering terms in (Goldberg 1989). This process simulates Darwinian evolution, which is considered the best natural model of the behavior of self-improving living organisms. Decision rules are considered to be a group of living pseudo-organisms, and their behavior follows the principle of survival of the fittest. The generation of rules is a multistage evolutionary process. At each stage of the process the system is supposed to produce improved rules, called offspring, in a way similar to biological reproduction. Individual rules are evaluated using different criteria. For example, a selection probability measures the probability that a given rule is correct when used against the body of test examples. The evolution of rules includes crossover (sexual reproduction), mutation (asexual reproduction), and inversion. Crossover occurs when two or more decision rules, or their parts, are combined. Mutation is a process of random changes of values of individual attributes in a given rule. Inversion is a change of dependent attributes into independent, and vice versa. There are proper genetic operators to conduct these three types of evolution.
There are four major stages in the operation of an inductive system based on genetic algorithms (Forsyth and Rada 1986):

1. Randomly generate an initial population of decision rules.
2. Evaluate these rules. If their overall average "quality" is good enough, halt and display the best of them.
3. For each rule, compute its evaluation score, i.e., its selection probability $p=e/E$ called evaluation score, where e is its individual score and E is the total score of all

rules.

4. Generate the next population of decision rules using selection probabilities and genetic operators. Repeat from stage 2.

SuperExpert

SuperExpert is a software package developed by Intelligent Terminals of Glasgow, United Kingdom. This system is based on the ID3 learning algorithm proposed by Quinlan (1986). SuperExpert produces a decision tree for the classification of examples into one of the decision attribute categories. There are four major stages in the operation of ID3:

1. Random selection of a window, a subset of examples from a given training set.

2. Use of the CLS algorithm to determine a classification rule for the assumed window for the division of the window examples into two or more decision attribute categories.

3. Search through the entire collection of examples to find exceptions to the latest rule.

4. If exceptions are found, add them to the window and repeat the process from stage 2. Otherwise, the process is stopped and the rules displayed.

The entire process of extracting decision rules from examples can be called "exception-driven filtering." CLS stands for Concept Learning System. It was proposed by Hunt (Hunt et al. 1966). It is based on the model of human simple concept formation, developed in experimental psychology in the 50's and 60's. The process of forming classification rules starts with the analysis of condition attributes. All attributes are compared according to their discriminatory power, i.e., their ability to partition data (examples in the window) into two categories of decision attribute. Quinlan proposed the use of entropy as an information-measure to quantify this power and to determine the most discriminatory attribute. This attribute is next used to divide the window into two subsets and to determine the root of the decision tree and the first decision rule. The initial two subsets are then considered separately and the same partitioning process is repeated for each of them. Each partitioning produces further nodes and decision rules, which are added to the decision tree. The process of partitioning is continued until no further partition of the subsets of examples is possible.

ID3 has several important limitations from the practical point of view. It produces only deterministic rules, i.e., rules that are always correct. When two examples are contradictory, the system eliminates them or signals clash (see Section 5). For these reasons the learning process is sensitive to small changes in the collection of training examples, and the preparation of examples and their consistency are quite important.

INDUCE

INDUCE is an experimental inductive system developed by Michalski (1983). It employs incremental multi-step learning. The final product of the learning process is a set of decision rules for the classification of given examples into one of the decision attribute categories.

At a given learning step all examples are divided into positive and negative. Positive examples are those classified into the category of the decision attribute under consideration while the remainder are negative. At each step another example is added to the examples being considered. The system extends the decision rule developed in the preceding stage by simply adding additional attributes and their values, or by building a star. A star is a complex decision rule containing descriptions linked by disjunctions ("or" symbol). A description is a group of attributes and their values connected by conjunctions ("and" symbol).

The system at first produces very specific decision rules which are gradually generalized. The initial set of examples can be considered as the initial collection of very specific decision rules, and each example is a description.

The generation of decision rules uses a four-stage process:

1. Randomly select a set of examples from the given training examples.

2. Maximally generalize each description of the selected examples while preserving the satisfaction of constraints.

3. Analyze descriptions and retain only the most general, nonredundant descriptions covering each concept.

4. If any description covers enough examples, display it.

DataQuest

DataQuest has been developed by Reduct Systems, Inc. of Regina, Saskatchewan, Canada. The system is based on a learning algorithm utilizing the theory of rough sets (Pawlak 1986). DataQuest produces a table with decision rules for the classification of examples into one of the decision attribute categories. The system has three major components: (1) data analysis, (2) rule generation, (3) consultation.

The system accepts data provided in attribute-value format in data tables, and prepared using a spreadsheet-like editor. Prior to the use of the system two groups of attributes must defined: condition, or independent attributes, and a single decision, or dependent, attribute.

The objective of data analysis is to identify and characterize the degree of functional relationship between condition and decision attributes. The system computes the degree of dependency between these two groups of attributes and allows the identification of dominant factors contributing to this dependency. Other data analysis routines allow for the elimination of noise, or irrelevant attributes, from the data before rule generation is attempted.

The rule generation component produces rules in a decision table format. The latest version of the system can process incomplete data, i.e., data with some unknown attribute values. In the case of nondeterministic data tables the system generates non-deterministic decision rules. These rules allow for more than one conclusion, or value of decision attribute, for a given combination of condition attribute values. The generated decision rules can be stored and used later during consultation.

The consultation component allows the user to use the generated decision rules to obtain a prediction of the decision attribute value for a given combination of condition attribute values. If a non-contradictory set of matching decision rules is encountered, the corresponding decision is displayed. Several alternative values of the decision attribute are produced in the case of nondeterministic decision rules. The strength of the prediction produced by the system is also calculated based on the degree of prediction support associated with each decision rule.

5. Comparison of Applications of Selected Inductive Systems

Knowledge Acquisition Processes Conducted.

The following three inductive systems have been used for knowledge acquisition: BEAGLE, DataQuest, and SuperExpert. The data was given in the form of examples. The objective of the comparison was to acquire knowledge from these examples. The acquired knowledge is in the form of decision rules (BEAGLE), a decision tree (SuperExpert), and a decision table (DataQuest). All three systems were used in the process of knowledge acquisition from the same collection of examples from the area of construction safety.

Construction safety is a complex domain of construction engineering, encompassing the analysis and prevention of construction accidents. The present understanding of the causal factors of such accidents is limited. There are neither general formal models nor any theory describing the occurrence and character of construction accidents. For these reasons construction safety experts are particularly interested in novel approaches to construction accident knowledge acquisition, which may improve the present understanding of accidents and their prevention. There is ongoing research on construction safety in the Intelligent Computers Laboratory of Wayne State University's Civil Engineering Department, and the results reported here are products of this research.

The examples used for our comparison were selected from a much larger collection of examples, based on records of past accidents provided by the National Institute for Occupational Safety and Health, and by a large construction company. These records involve the use of 21 nominal attributes, but for simplicity only five attributes and their values were used. The examples used in our experiments represent different cases of falls from elevations. The knowledge acquired from these experiments is in various forms, and should be considered only in the context of the small collection of examples used; obviously, any generalization of this knowledge to a wider domain would be inappropriate.

The attributes which have been used and their symbolic values are shown in Table 2, and the entire collection of examples is given in Table 3. The attribute BodyPart was selected as a dependent attribute. It identifies the body part injured in an accident. The Table 3 contains two identical examples and two clashing examples. Clashing examples are two or more examples having the same combination of condition attribute values but different decision or outcome attribute values.

Table 2. Construction Accidents, Falls: Selected
Attributes and their Values

Attributes	Values			
BodyPart	Back	Leg	Head	Neck
SurfStruck	Soil	HardSurf		
FallHeight	Low	Medium	High	
VictAge	Young	Medium	Old	
VictOccup	Carpenter	Plumber	SheetM-Work	Iron-Worker

In the case of BEAGLE, all nominal attributes were converted into numerical attributes. Although BEAGLE can perform analysis using both numerical and nominal attributes, it is not very efficient in dealing with nominal attributes of string values. It converts all character strings into numbers, and the interpreting the results may be quite inconvenient, particularly when fractional answers result. For this reason it is much better to use our own data conversion and avoid this interpretation problem. It should be noted that we conducted a double data conversion, introducing undesirable additional noise. The first conversion is from the original raw data in the accident records in Table 3, based on the use of nominal attributes. The second one is from nominal attributes to numbers. It would be wise to use the original raw data with BEAGLE instead of twice-converted data. In this case, however, different input data in terms of attributes and their values would be used in the knowledge acquisition processes conducted using BEAGLE and two other inductive systems. Thus, results obtained might be incompatible and difficult to compare.

Comparison of Results.

BEAGLE

BEAGLE conducts the learning processes separately for individual target expressions. A target expression is a combination of attributes and their assumed values; it is considered a predicate. The target expression is true when attributes are equal to their assumed values, and false otherwise. The system produces decision rules, expressed in terms of attributes not used in the target expression, for predicting the

Table 3. Construction Accidents, Falls: Collection of Examples

Example Number	Independent Attributes				Dependent Attribute
	SurfStruck	FallHeight	VictAge	VictOccup	BodyPart
1	HardSurf	High	Young	Carpenter	Back
2	Soil	Medium	Medium	Plumber	Leg
3	HardSurf	High	Young	SheetMWork	Head
4	HardSurf	High	Young	Plumber	Head
5	Soil	Medium	Medium	Carpenter	Head
6	HardSurf	High	Medium	IronWorker	Neck
7	HardSurf	Medium	Old	Plumber	Neck
8	HrdSurf	High	Old	Plumber	Head
9	HardSurf	Low	Medium	Carpenter	Back
10	HardSurf	Medium	Medium	SheetMWork	Neck
11	Soil	Hiegh	Young	SheetMWork	Head
12	Soil	High	Young	IronWorker	Head
13	HardSurf	High	Young	SheetMWork	Back
14	HardSurf	High	Old	Plumber	Head
15	HardSurf	High	Young	Carpenter	Neck
16	Soil	Low	Medium	IronWorker	Leg
17	HardSurf	Medium	Medium	Plumber	Back
18	Soil	High	Young	Plumber	Head
19	HardSurf	High	Old	Plumber	Head
20	HardSurf	Medium	Young	Carpenter	Neck
21	Hardsurf	High	Medium	SheetMWork	Head
22	HardSurf	Medium	Young	Plumber	Back
23	Soil	High	Medium	Plumber	Head
24	Soil	Medium	Young	SheetMWork	Back
25	HardSurf	Medium	Medium	IronWorker	Head
26	Soil	Medium	Medium	Plumber	Back
27	HardSurf	Low	Old	Carpenter	Leg
28	HardSurf	High	Medium	IronWorker	Head

state of the target expression. We assumed the target expression BodyPart = leg, since the other two systems in our study produced very simple decision rules for this target, thus simplifying the comparison between systems.

At the beginning of the learning process the user assumes an arbitrary division of examples into training and test sets. In our case 23 examples were assumed training examples and five test examples.

BEAGLE used these examples to generate randomly ten decision rules, to be used later as the initial population of rules. The first generated rule was:

if (FallHeight > VictAge) or (VictAge <= FallHeight) then BodyPart = leg

The rule has a quality score of 82.41. This roughly represents about 82 percent of the maximum possible departure from chance expectation. The quality score varies between 0 and 100, and any score above 60 is considered good.

The initial population of rules was then used in the evolutionary learning process, which was conducted assuming 10, 50, and 100 generations of decision rules respectively.

The first ten generations produced decision rules, of which the best one was:

if (VictAge < VictOccup) then BodyPart = leg

The quality score is 66, and this represents a worsening of results with respect to the initial population. Actually, the decision rule is quite general and can be written as three rules:

if VictOccup   = carpenter and VictAge = medium then BodyPart =   leg
if VictOccup   = carpenter and VictAge = old then BodyPart   =   leg
if VictOccup   = plumber and VictAge = old then BodyPart   =   leg.

When fifty generations were requested, BEAGLE terminated the learning process after only 38 generations. This was caused by the lack of progress in learning. The system produced the same decision rule as obtained after ten generations. Similarly, when we ordered 100 generations the learning process was stopped after sixty generations, and again the same decision rule was produced.

The entire learning process was repeated several times. It is known that BEAGLE sometimes produces poor results in one run and very good results in the next run due to the influence of randomness. Unfortunately, in our experiments the results were consistently similar to the ones described here. Most likely, more repetitions were necessary to produce meaningful results.

DataQuest

DataQuest requires the user to identify the decision (dependent) attribute and the group of condition (independent) attributes to be used in the learning process. The decision attribute has several states representing its different values. It is also necessary to specify whether decision rules related to all states of the decision attribute should be generated, or only those related to the selected states. In our case the decision attribute was BodyPart and all remaining attributes were used as condition attributes. All states of the decision attribute were considered.

DataQuest conducts its learning process in one-shot, and there was no need to divide the body of examples as in the case of BEAGLE. When the learning was conducted, it produced a decision table (Table 4). There are two decision rules related to leg injuries:

if SurfStr   = soil and Fall Heigh   = low then Body Part = leg
if VictAge   = old and Fall Heigh   = low then Body Part = leg.

Both decision rules are deterministic, i.e., they are always satisfied in the case of a given collection of examples. The system also produced four decision rules which are non-deterministic, i.e., there is more then one outcome associated with each rule. For example, the rule if FallHeight = High and VictOccup = Carpenter then BodyPart = Neck or Back. Nondeterministic rules reflect the lack of information in the input table for performing a complete discrimination of injuries based on predefined conditions. Nondeterministic rules are valuable from the practical point of view, as they provide an indication of the possible outcomes associated with different combinations of conditions. Such "weak" decision rules are also useful when we are want to learn about a given problem, or when its complexity and character justify the use of such decision rules in a knowledge-based system.

SuperExpert

The user of SuperExpert must identify attributes describing the problem, determine their type (logical or integer), and assume a class. The logical (nominal) attributes have symbolic values while integer attributes have numerical values. A class is a classification attribute which is always considered as logical. In our case Class was BodyPart, and all remaining attributes were logical.

Table 4. DataQuest: Decision Table.

Decision Rule Number	Independent Attributes				Dependent Attribute
	SurfStruck	FallHeight	VictAge	VictOccup	Bodypart
1		Medium	Young	Plumber	Back
2	HardSurf		Medium	Plumber	Back
3	HardSurf		Medium	Carpenter	Back
4		Medium	Young	SheetMWork	Back
5		Low	Old		Leg
6	Soil	Low			Leg
7		High	Medium	SheetMWork	Head
8		High		Plumber	Head
9		Medium		IronWorker	Head
10	Soil	High			Head
11	Soil			Carpenter	Head
12	HardSurf	Medium		Carpenter	Neck
13		Medium		SheetMWork	Neck
14		Medium	Old		Neck
15		High		Carpenter	Head or Back
16	HardSurf		Young	SheetMWork	Neck or Back
17	HardSurf	High		IronWorker	Neck or Head
18	Soil	Medium		Plumber	Back or Leg

Before the learning process is initiated the user decides whether testing of examples should be conducted. Testing reveals duplicate examples, which will simply be ignored, and clashing examples, if any, i.e., examples with identical attribute values but different class values. When testing is requested, all clashing examples are removed from the collection of examples and stored in the examples shelf, in a collection of inactive examples. We did not use the testing facility, and we expected to obtain results with some inconsistencies, reflecting possible example clashes.

The learning process produced the decision tree shown in Fig. 2. There are four clashes in this decision tree, corresponding to the nondeterministic decision rules obtained using DataQuest. For comparison purposes, two decision rules related to leg injuries are shown below:

if SurfStr   = soil and Fall Heigh   = low then BodyPart = leg
if VictAge   = old and FallHeigh   = low then BodyPart = leg.

These two decision rules are identical with decision rules obtained using DataQuest. For the two cases system could not produce any predictions of Body Part because of the lack of necessary examples. It was indicated by NULL values in the decision tree.

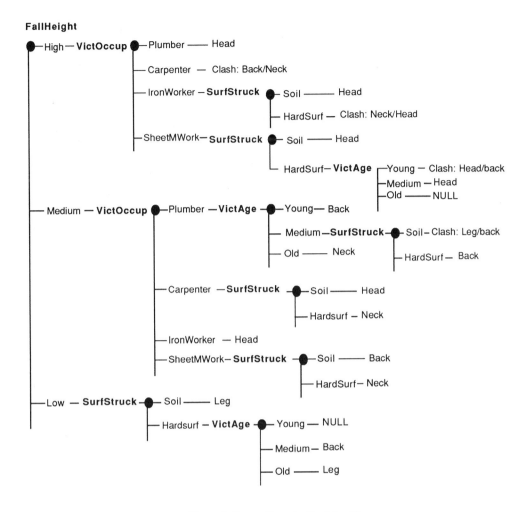

Figure 2. Super Expert: Decision Tree

6. Future Developments

Automated knowledge acquisition is still in a developing stage, and much of fundamental and applied research remains to be done. However, several commercial tools are now available, and can be used for both research and practical knowledge acquisition purposes. These tools demonstrate the potential of automated knowledge acquisition. Future developments in this area will be directly tied to progress in computer science and computer engineering, and should be considered in this context.

When comparing the state of computer science thirty years ago and now, one has to conclude that the majority of developments in this period were quantitative rather than qualitative in character. This means that, essentially, we are on the level of the 50's in terms of the basic concepts and ideas underlying computer technology. The basic structure of a computer system, composed of sequential addressable memory, registers, and CPU specialized to perform arithmetic operations, has not changed since the first computers were built. Computers are now much faster, the memories more capable, but almost everything is based on the old Von Neuman architecture. The technology of software engineering also has not made much progress since the 50's, chiefly because it is closely coupled to the standard architecture of a sequential computer. Parallel architecture and programming methods are still in an experimental stage, and it will probably take another ten to fifteen years before new parallel systems become widely popular. One limitation of past and current computers is particularly visible: the lack of the natural ability to "observe" the environment through sensors and to recognize patterns and regularities in observed data for the purpose of recognition, reasoning, prediction, control, or knowledge acquisition. Current computers are unable to deal with unstructured "fuzzy" input data such as sound, images, etc. They are not capable of learning from the observation by themselves, and they are unable to discover. The keyboard is still the primary means of communication with the external world and, by its nature, supplies very limited amounts of usually well structured, clean data. Only special-purpose systems are equipped with analog-to-digital convertors which can supply floods of unstructured raw data from a variety of sensors such as thermometers, microphones, cameras, etc. Current machines, unfortunately, are not able to sort out this kind of data in an intelligent way, or to discover essential data relationships or regularities for automatically forming and improving an internal model of the real world. The efforts of researchers in the areas of pattern recognition, machine learning, machine discovery, fuzzy and rough sets, and neural nets, are aimed at developing methodologies for use as a basis for future software to control the machine learning and discovery processes. A "discovery" software would be an internal part of a computer system, as operating systems are today. The computer system would be equipped with a variety of sensors (sound, pressure, image, etc.) and would automatically learn some of the things that humans learn, such as recognizing voices, images, smells, and so on. The internal "discovery" software would continuously sense the environment and reprogram itself dynamically to perfect its internal recognition model. This type of behavior, however, will not be achieved by increasing the speed and capacity of computers. Fundamentally new ideas will have to be developed and implemented in the areas of basic mathematics, computer principles, and software engineering. Optimistically, it will require another ten to twenty years before such truly intelligent systems with automated knowledge acquisition capabilities emerge as popular computers. Special-purpose systems of this kind probably will be available within the next five to ten years, serving industrial installations requiring adaptive control based on "fuzzy pattern" recognition capabilities.

Appendix-References

Adeli, H., Yeah, C., (1990). "Explanation Based Learning", *Engineering Applications of Artificial Intelligence*, Vol.3.

Adeli, H., Yeah, C. (1989). "Perceptron Learning in Engineering Design", *Microcomputers in Civil Engineering*, No. 4, pp 247-256.

Adeli, H., (edit.), (1989). *Knowledge Engineering*, Vol. 1,2, McGraw-Hill, New York.

Arciszewski T., Mustafa M., (1989). "Inductive Learning Process: The User's Perspective," chapter, in the book *Machine Learning*, edited by R. Forsyth, Chapman and Hall, London.

Arciszewski, T., Ziarko, W., (1990). "Inductive Learning in Civil Engineering: Rough Sets Approach," *Microcomputers in Civil Engineering*, No. 3.

Arciszewski T., Ziarko W., (1987). "Adaptive Expert System for Preliminary Engineering Design," *Revue Internationale de CFAO et D'Intographie*, Vol. 2, No. 1.

Arciszewski T., Mustafa M., Ziarko W., (1987) "A Methodology of Design Knowledge Acquisition for Use in Learning Expert Systems," *Int. J. of Man-Machine Studies*, No. 27.

Barr A., Feigenbaum, E.A., (1981). *The Handbook of Artificial Intelligence*, Vol. 1, William Kaufman Inc, Los Altos, California.

Burstein, M.H., (1986). "Analogical Learning with Multiple Models," *Machine Learning: A Guide to Current Research*, ed. Mitchell T.M., Carbonell, J.G., Michalski, R.S., Kluwer Academic Publishers, Boston.

Cullen J., Bryman A., (1988). The Knowledge Acquisition Bottleneck: Time for Reassessment,? *Expert Systems*, Vol. 5, No.3, pp. 216-224.

Firebaugh, M.W. (1988). *Artificial Intelligence: A Knowledge-Based Approach*, Boyd and Fraser Publishing.

Forsyth, R., (1989) "*Machine Learning, Principles and Techniques*, (edit.), Forsyth, R., Chapman and Hall Computing, London.

Forsyth, R., Rada, R., (1986) *Machine Learning, Applications in Expert Systems and Information Retrieval*, Ellis Horwood Limited, Chichester, England.

Harandi, M.,T., Lange, R., (1990) "Model-Based Knowledge Acquisition," in *Knowledge Engineering*, Vol. I, edit. Adeli, H., McGraw-Hill, New York.

Holland J.H., Holyoak K.J., Nisbett R.E., Thagord P.R., (1987). *Induction: Process of Inference, Learning, and Discovery*, MIT Press, Cambridge, Massachusetts.

Hunt, E.B., (1986). Marin, J., Stone, D., *Experiments in Induction*, Academic Press, London.

Holland J.H., (1975) "*Adaptation in Natural and Artificial Systems*," University of Michigan Press, Ann Arbor, Michigan.

Goldberg D.E., (1989). *Genetic Algorithms in Search, Optimization and Machine Learning*, Addison-Wesley Publishing Company, Reading, Massachusetts.

Intelligent Terminals, (1986) *Super Expert, Manual*, Softsync, Inc, New York.

Kedar-Cabelli, S., (1986). "Purpose-Directed Analogy, A Summery of Current Research," *Machine Learning: A Guide to Current Research*, ed. Mitchell T.M., Carbonell, J.G., Michalski, R.S., Kluwer Academic Publishers, Boston.

Kondratoff, Y., (1990). "Machine Learning," in *Knowledge Engineering*, Vol. I, edit. Adeli, H., McGraw-Hill, New York.

Lenat, D.B., (1983). "The Role of Heuristicsc in Learning from Observation and Discovery: Three Case Studies," in *Machine Learning: An Artificial Intelligence Approach*, Michalski R.S., Carbonell J.G., And Mitchell T.M., (eds), Morgan Kaufman Publishers, Los Altos, California.

Mustafa, M., Arciszewski, T., (1989). "Knowledge Acquisition: Engineering Methodology of Inductive Learning", *Proceedings of Workshop on Knowledge Acquisition, Joint International Conference on Artificial Intelligence*, Detroit.

McGraw K.L., Harbison-Briggs K., (1989). *Knowledge Acquisition, Principles and Guidelines*, Prentice Hall, Englewood Clitts, New York.

Michalski R.S., Carbonell J.G., and Mitchell T.M., (eds), (1983). *Machine Learning: An Artificial Intelligence Approach*, Morgan Kaufman Publishers, Los Altos, California.

Minsky, U. (1973). "A Framework for Representing Knowledge," Winston P., (ed.)," *The Psychology of Computer Vision*, McGraw-Hill, New York.

Mostow, D.J. (1983). "Machine Transformation of Advice into a Heuristic Search Procedure," in *Machine Learning: An Artificial Intelligence Approach*, Michalski R.S., Carbonell J.G., and Mitchell T.M., (eds), Morgan Kaufman Publishers, Los Altos, California.

Nillson N.J., (1965) *Learning Machines: Foundations of Trainable Pattern-Classifying Systems*, McGraw-Hill , New York.

Nilsson, N.J. (1971). *Problem-Solving Methods in Artificial Intelligence*, McGraw-Hill, New York.

Pawlak, A., (1982). Rough Sets, *Intl. J. of Computer and Information Sciences*, Academic Press Limited, Vol 5, No. 11, pp. 341-356.

Popper, K., (1959). *The Logic of Scientific Discovery*, Basic Books, New York.

Reduct Systems Inc., (1991). *DataQuest - User Manual*, Regina, Canada.

Rosenblatt F., (1987) "The Perceptron: A Perceiving and Recognizing Automation," *Project PARA*, Cornell Aeronautical Laboratory Report 85-460-1.

Rumelhart D., Hinton G., Williams R., "Learning Internal Representations by Error Propagation", in *Distributed Processing: Explorations in the Microstructure of Cognition*, Rumelhart, D.E., And McCleveland J.L. (eds.) MIT Press, Cambridge, Massachusets.

Samuel A.L. (1959) "Some Studies in Machine Learning Using the Game of Checkers," IBM. *J. of Research and Development*, No.3.

Simon H.A., (1983) "Why Should Machines Learn?", in *Machine Learning: An Artificial Intelligence Approach*, Michalski R.S., Carbonell J.G., And Mitchell T.M., (eds), Morgan Kaufman Publishers, Los Altos, California.

Turing, A., (1950) "Computing Machinery and Intelligence," *Mind*, No. 50.

Quinlan R., (1983) "Learning Efficient Classification Procedures and Their Applications to Chess End-Games," in *Machine Learning: An Artificial Intelligence Approach*, Michalski R.S., Carbonell J.G., Mitchell T.M., (eds), Morgan Kaufman Publishers, Los Altos, California.

Quinlan, R.,(1986). "Induction of Decision Trees," *J. Machine Learning*, Kluwer Academic Publishers, Vol. 1, No. 1, pp. 81-106, Boston.

Warm Boot Ltd., (1987). *PC/BEAGLE, User Guide*, Nottingham, United Kingdom.

Winston, P.H., (1977). *Artificial Intelligence*, Addison Wesley, Reading, Massachusetts.

Woods, W.A. (1968) "Procedural Semantics for a Question - Answering Machine," Proc. of the Fall Joint AAAI, Fall Joint Conference 33, 457-471.

Wong, S.K.M., Ziarko, W., (1986). "INFER - An Adaptive Decision Support System," *Proc. of the 6th Intl. Workshop on Expert Systems and Their Applications*, Avignon, France.

Ziarko, W., (1989). "Data Analysis and Case-Based Expert System Development Tool ROUGH," *Proc. of Case-Based Reasoning Workshop*, Pensacola Beach, Florida.

CHAPTER 4

**Knowledge Elicitation Strategies and Experiments
Applied to Construction**

Jesus M. De La Garza and C. William Ibbs

1. Introduction

This chapter focuses on the process of extracting the knowledge for a system, called CRITEX, written for the Corps of Engineers with the purpose of critiquing construction schedules. Its perspective is that of an owner's reviewing initial and in-progress schedules. This study is part of a considerable amount of research which is underway in applying Artificial Intelligence (AI) concepts, specifically Knowledge-Based Expert Systems (KBES), to project management systems (PMS) (Ashley and Levitt 1988).

Several investigators have argued that no single methodology for the process of knowledge elicitation has proven universally effective (Hart 1985; Hoffman 1987; Trimble et al. 1986; Trimble 1987). Our investigation was aimed at testing several different methods and identifying which construction-related circumstances favor each of these methods. The knowledge elicitation techniques utilized are: (1) published material review; (2) structured interview; and (3) observation of tasks and behavior with limited information. A construction schedule is the product of inputs from many people. Thus, it was important to find ways to fuse knowledge, sometimes expressed in contradictions, from several contributors. Mittal addresses this issue in an important article (Mittal 1985), and this research employed some of those concepts.

Mittal proceeds in that same article to argue that the advantages of using a diverse group of experts may well offset the extra time required to resolve conflicts generated by multiple inputs. Contradictions may in fact be turned into advantages as they indicate areas for further research. Sources of expertise for CRITEX came from four representatives: (1) a general building contractor; (2) an owner's resident engineer; (3) a scheduling consultant; and (4) university professors with scheduling experience. Conflicts did arise and had to be resolved by the researchers. An open review, consensus-forming type of technique was employed.

2. Background

The construction schedule critiquing domain is characterized by specific uses and key traits of expert knowledge, judgment and experience. Management of dynamic schedules requires knowing not only what has changed, but also reacting to those changes (if possible) before their impacts occur. In practice, schedules are typically compiled and critiqued at three different stages: (1) prior to the start of construction when the initial schedule version is (iteratively) completed; (2) during the project execution as the schedule is continually

69

updated; and (3) at the completion of the project, diagnosing the meanings of any deviations between baseline targets and actual results.

Computer-based schedule analysis for project management dates back to 1959 when program evaluation and review technique (PERT) and critical path method (CPM) were first reported. A study fifteen years later on the use of network-based techniques revealed that CPM and PERT were being employed principally for planning, rather than control as intended (Davis 1974). A key explanation for this unacceptance was the difficulty people had interacting with the systems. Complex input/output notation, excessive detail and inflexible report generation were reported to be at the root of this difficulty.

Since this survey was conducted, the increased availability and affordability of personal computing resources has led to a proliferation of computerized PMSs. After more than 20 years of commercial availability, however, PMSs are still acknowledged as incomplete project control aids. Cited deficiencies include their inabilities to: (1) collect and interpret qualitative or subjective project information; (2) be customized to a firm's idiosyncrasies; and (3) provide substantial diagnostic functions even of strictly quantitative data (Teicholz 1989). The underlying premise of this study is that the absence of such computer-aided interpretation features contributes to highly variable manager and firm scheduling effectiveness. The specific intent is to synthesize the operational knowledge needed to perform scheduling criticism from an owner's viewpoint (De La Garza and Ibbs 1988). In doing so we hope to show that AI concepts have validity in addressing at least some of the PMS deficiencies cited earlier.

Schedule criticism conducted by owners or contractors starts with validating the initial plan. Project managers and schedulers need to answer such questions as:

* Does the schedule meet the contract requirements?
* Is the critical path(s) reasonable?
* Are owner-controlled activities included?
* Have major subcontractors participated in the formulation of the plan?
* What is the overall degree of schedule criticality?
* Do procurement activities precede special installation tasks?
* Does the cost estimate comply with the contract documents?
* Have weather-sensitive activities been scheduled with likely weather conditions in mind?

A few of these examples may also be appropriate for in-progress evaluations. Subcontractor involvement is desirable throughout the job, for instance, as long as that subcontractor still has some relevance to the project. Once that subcontractor's association with the schedule is finished though, in-progress evaluations can downgrade or eliminate that party's input. In this mode, project managers face some of the same questions as in the initial analysis--perhaps with different degrees of importance--and other issues that did not arise in the original schedule analysis stage:

* Are we on schedule?
* How much money should we pay or collect in a given period?
* What is different about these schedules?
* Should the duration of future activities be modified based on past experience? If so, by how much?
* How can I tell if activities are in trouble?

This study was limited to critiquing original and in-progress schedules for medium-height commercial building projects because of research sponsor-interests and constraints. This particular focus has several advantages: (1) it represents a large segment of the total

Construction Industry; (2) it is characterized by repetitive operations; and (3) solutions to problems diagnosed early may be usefully propagated to the rest of the structure because of this repetition.

3. Knowledge Acquisition Process

This chapter focuses on construction scheduling knowledge elicitation, which is just one part of the entire knowledge metamorphosis (Fig. 1). The four stages of this metamorphosis are discussed briefly in the following paragraphs to put the knowledge elicitation discussion into context. The entire knowledge engineering process is analogous to a work breakdown approach. High-level broad concepts are repeatedly refined until some descriptions and actions are precisely specified.

Amorphous Knowledge Base. Amorphous knowledge is first stage information, such as facts and wisdom, which is collected from any source and which has no immediately discernible patterns or rationale behind it. Most information is, in fact, amorphous by this definition until it is reduced to a more traditional, computerized-form of data. "Pour large concrete slabs in subsections" is amorphous; while "Subdivide any concrete slab into equal partitions not exceeding 5000 square feet each before pouring" is more precise. Much construction common sense knowledge is never expressed in terms more precise than what are conveyed at this stage, which turns out to be a considerable obstacle to the widespread implementation of KBESs.

English Knowledge Base. CRITEX's English knowledge base consists of a set of conceptual scheduling provisions and specific procedures written in plain English. A conceptual schedule provision represents a generalized scheduling statement that has not yet been coupled with an interpretation. A specific procedure corresponds to one of several possible interpretations of a given conceptual schedule provision. Fig. 2 shows the knowledge elicitation techniques used in this research and their output. The overlapping depicted in the provisions and methods boxes symbolizes some duplication in their contents.

The union of all provisions defines the breadth of the knowledge base. An important attribute of these conceptual provisions is that they represent the "what" and do not explicitly indicate "how" they should be interpreted or implemented. For example, a conceptual scheduling provision in the "time" category basically states: "Float should be broad enough to support the premise that it has not been manipulated." In this case, the intent of the provision is not yet tied to any float sequestering technique.

Taken together, the procedures delineate the depth of the knowledge base. At this point in the decomposition process these procedures begin to define "how" a conceptual provision may be interpreted. The time-based conceptual provision stated above may be interpreted by relating it to preferential scheduling or float sequestering techniques. Examples would be preferential logic ties, lead/lag activities, or extended activity durations.

English-like Knowledge Base. Much construction scheduling knowledge is procedural. For instance, "Do the following in this exact manner." The main objective at this point of our knowledge engineering process was to transform conceptual scheduling provisions into executable procedures. Some project management knowledge is generic, and some is idiosyncratic to a company, which means there can be several ways to implement a conceptual scheduling provision, depending upon company practices.

Once a procedure is attached to a provision in this knowledge metamorphosis process, its representation is no longer in plain English. Instead it uses a less idiomatic and more procedural notation. A person unfamiliar with the specifics of the KBES computer language can still read and record additional knowledge.

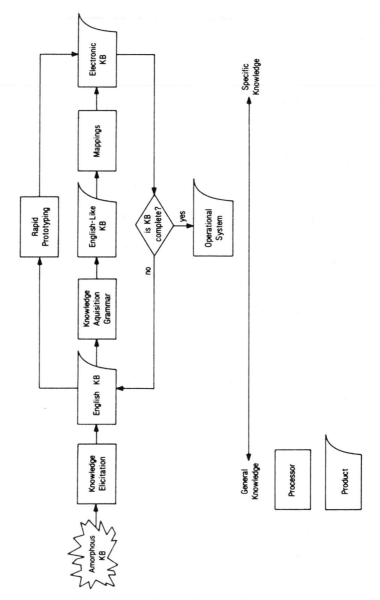

Figure 1. Knowledge Metamorphosis

Electronic Knowledge Base. Computer implementation of conceptual scheduling provisions requires the selection of programming languages. Commercially available AI programming environments, expert system shells and programming languages can be used. An AI programming facility like the automated reasoning tool (ART^{Tm}) requires a set of mappings to transform the knowledge written in the English-like notation into the specifics of the programming language syntax. Once in this format modifications to the knowledge base require programmer expertise or a high-level knowledge base interface.

The three knowledge acquisition techniques used in this work, and illustrated in Fig. 2, were adapted from Freiling et al. (1985), Kneale (1986), Michie (1986), and Prerau (1987). The following sections provide details about their application in this study.

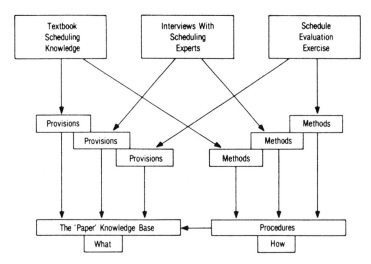

Figure 2. Knowledge Elicitation Techniques

Textbook Scheduling Knowledge. Our first knowledge elicitation methodology was an analysis of textbook scheduling knowledge. This step was used to explore the breadth and depth of the construction schedule criticism domain. This effort substantiated the notion that the initial and in-progress schedule analyses, as defined in this study, were sufficiently narrow and self-contained for application of KBES technology. Use of this approach also helped identify early on some key scheduling concepts that would be refined later by the other two knowledge acquisition techniques. These concepts were expressed as a categorized, plain-English list of dos and don'ts.

This first pass at developing a knowledge base consisted of analyzing available texts and technical manuals on the subject (Avots 1985; Clough 1981; Gray 1986; Levitt and Kunz 1985; O'Connor et al. 1982; Ponce de Leon 1984). Fig. 3 schematically illustrates the evolution of public domain know-how into human say-how. The analysis process helped us structure the knowledge base into four categories: (1) general requirements; (2) logic; (3) cost; and (4) time. Know-how in this diagram refers to the knowledge a person or computer possesses. The term say-how is used for that knowledge a person can articulate in plain English. Show-how refers to the actual demonstrated use of knowledge by a person or computer. They may differ from each other either blatantly or subtly.

Interviews With Schedulers. The second knowledge acquisition technique was a direct, structured interview of experienced schedulers. The knowledge base produced in the first phase influenced the interview structure. These scheduling professionals helped us to understand the different kinds of expertise prevalent in the domain, and to further define the four different rule categories of CRITEX. In addition, these schedulers were better able to understand the scope and complexity of our research. The articulation or dialogue acquisition technique was utilized to elicit and formalize the experts' knowledge. This method is schematically depicted in Fig. 3. Its purpose was to convert human know-how into human say-how through a process of articulation by the experts.

The articulation channel of Fig. 3 contains what AI researchers call the knowledge acquisition problem, also known as the "Feigenbaum bottleneck" (Feigenbaum and McCorduck 1983). Articulating knowledge is a complicated process because experts are expected to express their knowledge perfectly and the experts may not have enough

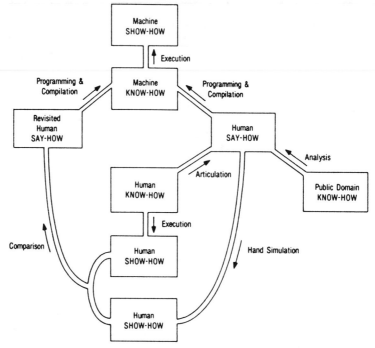

Figure 3. Analysis, Dialogue, and Execution Route Maps

patience to detail what they might consider common sense issues. The narrowness of the bottleneck is directly related to task domain complexity. Michie (1986) argues that the more complex the mental skill, the greater the proportion of it which is encoded in intuitive form and hence tends to be more difficult to explain.

For this research, two building contractor schedulers were interviewed separately (one for twelve hours and another for four hours). Sixteen hours' of interviewing were equally divided into four sessions. This structured interview technique was utilized mainly because these senior project personnel were reasonably expected to be capable of expressing the large body of pattern-based rules which they mentally possess. These schedulers reviewed the first-generation knowledge base one entry at a time, making comments on each. This process led to addition, modification and deletion of entries, quantification of entries, and even addition of broad categories of entries. It led to a definite improvement in knowledge base quality.

The combination of these two knowledge elicitation approaches produced a more complete series of high-level concepts; i.e. a second pass to human say-how. However, the two schedulers often could not describe just how much they knew, how they kept track of it, or how they knew when to use which information. In this light, a less direct approach to knowledge acquisition was conducted.

Scheduling Evaluation Exercise. The third technique involved observing human performance on specific problems. Limiting the information which is typically available to the expert schedulers and constraining the problem they were to work on made it possible to stimulate intuitive knowledge.

This third knowledge acquisition alternative can be called observation of task performance under limited information constraints. It assumes that there is some other way of going from human know-how to machine know-how. As shown in Fig. 3, it is possible to proceed from human know-how into tutorial human show-how by stimulating expert knowledge and reasoning skills. Moreover, in the development of CRITEX it was possible to go from the second-generation human say-how to human show-how by hand-simulating the application of this knowledge base. After arriving at this point, a model of the experts' skill in explicit form was translated into third-generation human say-how rules. Some of those rules have been subsequently compiled into machine know-how.

This knowledge acquisition approach largely circumvents the articulation bottleneck, a powerful reason to employ it. The method does consume extensive time and effort on everyone's part, though it is not clear that more effort is needed with this approach than with the structured interview process.

The critical processes for implementing this approach are represented in Fig. 3 by the bridges that link the human know-how and say-how with the human show-how. These processes are the execution and hand-simulation channels. To simulate these channels, an experiment was designed. It included two video cameras, a trio of senior project managers, a rookie project manager, and a curtain. The aim was to mimic a computer by having the trio act as the KBES, the curtain act as the computer screen and the rookie act as the user (Kneale 1986).

This exercise had four goals: (1) expand the knowledge base toward completeness by eliciting new knowledge; (2) validate, fine tune and, if necessary, disprove existing rules in the knowledge base; (3) develop a model of interaction between the user and the knowledge-based computer system; and (4) determine the feasibility of this technique to capture actual construction scheduling knowledge.

To keep control on the information exchange between the rookie and the experts, the actual setting consisted of two separate rooms. In one room the rookie conducted a construction schedule audit. He only used the reports generated by two PMSs and the advice coming from the trio of experienced schedulers. These PMS reports contained different sorts of data associated with each schedule activity, such as early and late start dates, float, cost, successors and predecessors, duration, and description.

While the video cameras recorded the approach, the experts, located in a separate room, guided the rookie in his schedule criticism task. The question/answer communication protocol between the rookie and the experts was implemented in electronic mail messages on a Local Area Network (LAN).

Schedule information available to the trio was restricted. They had only a tablet and a pencil with which to work. The hypothesis was that if project schedule information is withheld, the experts would have to rely heavily upon and provide additional evidence about their knowledge and reasoning skills (Hoffman 1987; Kneale 1986). By scrutinizing their tactics and procedures we were soon able to anticipate actually what data they wanted, and the type and sequence of information that they produced.

While the trio and rookie interacted, an attempt was also made to use the existing rules and procedures derived from the previous two knowledge elicitation techniques. A fourth individual acted as process monitor and assessor. This hand-simulation process provided us with the opportunity to examine the usefulness and completeness of the second-generation knowledge base simultaneously. This comparison identified points of disagreement and consensus, and generated a more solid human say-how. In general, this effort reaped substantial gains to the knowledge base.

A limited two-day exercise was conducted with a contractor, an owner and a PMS developer. An introductory period was used to present an overview of KBES, the objectives of this research project, a detailed description of the exercise mechanics, and a general statement regarding their expected contribution. They also had the opportunity to get acquainted with the Banyan[Tm] LAN. One experimental variable that we tried to assess was the level of interaction among the experts. This was difficult and we did not have much success other than to conclude that this interaction is important to idea generation and conflict resolution.

During the 2-day exercise four distinct phases were set aside for the research, corresponding to our information needs:

i)   The initial schedule criticism phase. During this phase, the rookie had to criticize the schedule of a low-rise medical dispensary building. The rookie had access to general project characteristics, scheduling notation, number of activities, job location, cost, start and finish time, and the building's structural design system. Three hours were allocated to this phase.

ii)   The in-progress schedule criticism phase. During this phase, in-progress scheduling practices were explored. The building project was slightly different from that used in the initial scheduling practice review. The reason for this change was primarily to present the participants with a fresh project, to avoid boredom. Three hours were assigned to this stage.

iii) The overall schedule criticism phase. The purpose of this phase was to address automatic schedule generation issues as well as schedule logic analysis. No actual construction project was referenced here.   Rather, the parties conversed electronically only at a conceptual scheduling level. Two hours were devoted to this period.

iv)   The de-briefing phase. This knowledge elicitation technique is relatively untested so we felt that it was especially important to de-brief the trio and rookie. This feedback will be available to improve future knowledge acquisition efforts. De-briefing lasted one hour.

CRITEX's paper knowledge base consists of a set of conceptual schedule provisions written in plain English. This knowledge is contained within the component labeled "English KB" in Fig. 1. These provisions are the result of analyzing and synthesizing knowledge produced by the three knowledge acquisition techniques (textbook reviews, personal interviews and the evaluation exercise). They represent the elements (or the "what") with which construction schedules need to comply. Generally, they do not provide details as to "how" implementation should be handled.

Scheduling principles represented by the paper knowledge base are conceptual in nature by our approach and not limited to a single interpretation. Therefore, they are applicable to a wider audience of owners and contractors. These synthesized construction scheduling principles can then be used as guidelines for performing construction schedule criticism and be modified to suit the idiosyncrasies of the individual user.

The Abstraction Process. A retrospective analysis of the intuitive synthesization process which produced the majority of the conceptual schedule provisions is outlined in this subsection. For a given set of examples or versions of the same scheduling method the following algorithm, broken down into steps, was used:

Step 1: For each entry in the set of examples find an underlying scheduling principle which is compatible with previous generalizations of the same scheduling method and that includes the example under consideration. Repeat the process until all examples in the set are covered.

Step 2: If the synthesized scheduling provision of step one does not lend itself to adaptation by some other firms which did not participate in this study, then continue its generalization. Repeat the process until there is an indication that the scheduling principle is general enough that companies not involved in the study might also use it.

To illustrate the application of the synthesization process, the generation of two schedule provisions is presented next.

Provision #1: The three knowledge acquisition techniques contributed four examples which correspond to different methods of defining activities durations. They are:

i) Activity duration limits usually range from one to 30 calendar days.
ii) Reasonable activity durations range from five to 25 days.
iii) Critical activities should have enough detail; that is, have a duration of less than 20 days.
iv) A typical activity duration should be between five and 25 days.

The application of step one in the abstraction process corresponds to the following four iterations:

i) Activities durations need to be constrained by a maximum duration.
ii) Activities durations need to be constrained by a minimum-maximum range.
iii) Same as second.
iv) Same as second.

The application of step two in the abstraction process to further generalize the scheduling provision of step one corresponds to the actual schedule provision, namely: All activities affecting progress should be included and defined in a way that they can be easily monitored and measured. The total number of activities should remain manageable.

Provision #2: The three knowledge acquisition techniques contributed five examples which correspond to different methods of predicting significant deviations from the official schedule. They are:

i) Analyze activities whose remaining durations exceed 20 days.
ii) Ensure that the ratio of (days to makeup/total days remaining) is less than 0.20.
iii) If the contractor needs to extend the duration of critical, unfinished activities based on the progress of the project to this point, the contractor will also modify the logic of other activities to compensate for such delays.
iv) If the activity variance report shows the project ahead of schedule, the contractor will notify the owner of such gain; however, the contractor will not commit to some future unrealized gain.
v) The contractor will be reluctant to change the logic of unfinished activities even if the project is ahead of schedule.

The application of step one in the abstraction process to methods one and two corresponds to the following first generalization:

i) Identify areas that need remedial action while there is still time for such action to produce a positive effect.

The application of step one in the abstraction process to methods three, four, and five corresponds to the following, second generalization:

ii) Determine whether unfinished and finished activities have anything in common, e.g., subcontractor, material type, crew, etc.

The application of step two in the abstraction process to further generalize the scheduling provisions of step one corresponds to the actual schedule provision, namely: Schedule projections should be based on comparisons between what was planned and what actually happens.

Examples of Conceptual Schedule Provisions. The knowledge base addresses four major rule categories, as mentioned previously: (1) General Requirements; (2) Logic; (3) Cost; and (4) Time. A scheduling provision example from each category illustrative of the plain English vernacular is provided here:

General Requirements. All activities affecting progress should be included and defined in a way that they can be easily monitored and measured. The total number of activities should remain manageable.

Time. Schedule projections should be based on comparisons between what was planned and what actually happens.

Cost. Cash flow front-end loading should be avoided.

Logic. An activity that is weather sensitive and not weather-protected should not be scheduled in a period when weather conditions are expected to be below specified minimums or above specified maximums.

Customization of Conceptual Schedule Provisions. AI researchers and users agree that there is nothing magic about KBESs. They generate conclusions that are already explicit or implicit in their knowledge bases. A KBES, however, cannot be built if we do not know or understand the task for which it is designed. This subsection shows how a knowledge base conceptual schedule provision may be transformed into executable procedures.

These procedures correspond to the expertise that is utilized by a firm. They represent one firm's way of performing schedule criticism. The implication is that the interpretation of one knowledge base conceptual schedule provision may be different from one firm to another. The differences in interpretation may just be the degree of sophistication, complexity, or detail that is given to a provision. Achieving homogeneous interpretations across all firms is not as important as having such interpretations concur with the original intent of the conceptual schedule provision. This flexibility in interpretations is consistent with the idea that a company will not utilize a KBES whose knowledge base represents another company's way of work.

Practically, it is extremely difficult to investigate and exhaust all possible ways of interpreting a conceptual schedule provision. Thus, only a few interpretations of one such conceptual schedule provision are presented here.

Provision: Cash flow front-end loading should be avoided. There are two basic questions that deserve consideration: (1) what are the possible ways to detect front-end loading in a schedule; and (2) how to discourage contractors from front-end loading construction schedules.

The following examples correspond to scenarios which could be utilized to detect the presence of cash flow front-end loading.

i)   Calculate the cumulative cash flow "S" curve. For mid-rise buildings it would show 1/4 of the cost committed at the end of the first third of the project, and 3/4 of the cost at the end of the second third. This measure should be applied with caution. The fact that the cost committed at 1/3 and 2/3 points of the project schedule is greater than 1/4 and 3/4 of the project cost, respectively, also is an indication that the last 1/3 of the schedule is too conservative and that the preceding 2/3 of the project schedule is too optimistic.

ii) Assess whether the mobilization costs are reasonable given the contractor's preliminary equipment list.

iii) Determine if there are items that are likely to overrun quantities. If so, assess whether their costs appear reasonable in relation to similar items with larger quantities.

iv) Examine the unit prices of early activities to determine if they are greater than unit prices of similar activities scheduled towards the end of the project.

v) Determine the amount of work subcontracted. The higher the amount of such work the less likely there is front-end loading.

There are, on the other hand, specific mechanisms that owners can use to dissuade contractors from front-end loading construction schedules. The following examples provide broad guidelines to accomplish this.

i) The monetary aspect of the bid should be determined on a Net Present Value (NPV) basis. This practice heavily penalizes front-end loading. However, it should be noted that it also discourages early completion of non-critical activities.
ii) Provide a financial investment mechanism, e.g., US Treasury Bonds, so that retainage earns interest.
iii) The cost of expensive materials and equipment should be separated from their installation cost.

4. Conclusions

This knowledge elicitation study has contributed additional insight into the process of extracting, articulating, and synthesizing construction schedule knowledge. The work has shown that the three knowledge engineering approaches used here are sound for the construction scheduling domain. This research also represents a modest contribution to the improvement of schedule management automation.

Generation and criticism of construction schedules is a process that is partly based on art and partly on engineering principles. Arguments can be offered as to which is dominant, although it is not important to determine their respective contributions precisely. Instead, what is relevant is that their presence carries some implications concerning the kind (empirical versus causal) and type (generic versus idiosyncratic) of knowledge that constitutes the scheduling process.

The scheduling knowledge elicited in this study is mostly empirical and idiosyncratic. It is empirical in that it involves practical expertise, although not necessarily with an underlying knowledge of causality. It is idiosyncratic in that it reproduces the knowledge and problem-solving style of the experts consulted. After all, if everything is held constant, a different but equally good schedule may be produced by another contractor. This point strongly suggests that for an operational KBES to be used by a firm, the KBES should reproduce quite accurately that company's natural way of doing things, which then indicates a need for more user participation in building construction scheduling KBESs.

There are 34 conceptual schedule provisions in CRITEX. No attempt is made in this study to precisely determine the boundaries of construction schedule criticism. Whether there are 34 or 1000 provisions is not accurately known. What the writers offer in this regard are 34 provisions that represent a large body of qualitative construction scheduling knowledge currently missing in commercial PMSs, making it worth developing a KBES.

Scheduling Evaluation Exercise Mechanics. Retrospective analysis of this experiment leads to the following conclusions.

i) The experts were at times bothered with the need to request every single piece of information they wanted. They would have preferred a more detailed, introductory, narrative description of the project to assimilate as much implicit detail as they could. This would have defeated the purpose of using the limited information approach, however, and the trio would not have revealed all the facts and relationships they processed in the critique.

ii)   The experts were frustrated by the number of calculations they had to perform by hand. These calculations included such heuristics as counting the number of simultaneous critical paths, and computing the amount of total and free float across all activities and similar sub-groups. The trio felt they were expected to immediately and completely absorb both explicit and implicit information.

iii) The roles of the trio and rookie were dynamic. While the trio was trying to obtain relevant information from the rookie, he sometimes was asking them questions to understand their process better. This circularity actually caused confusion a few times.

iv)   Sections of PMS reports were intentionally given to the professional schedulers to point out the disadvantages of using nonlimited information approaches. For example, when the trio had a PMS report containing critical activities with basic parameters, they spent considerable time silently analyzing the report. During these periods the trio's thought processes were being stimulated but not captured. Answers were being missed for questions like what kind of data relationships were they looking for? Why? What information was relevant and irrelevant?

v)   There was not always a one-to-one correspondence between the trio's advice and the information supplied by the rookie. Advice sometimes was given on the basis of the trio understanding a series of questions or project aspects contextually. This made it difficult, sometimes even impossible, to relate advice to individual project information requests. In such cases, the rookie repeatedly had to ask "why" and "how". Advice would also come in spurts.

vi)   At times, requests for elaboration became so detailed that the trio would characterize them as trivial, too demanding, or even dumb. When they became sufficiently frustrated, they could not articulate their knowledge patterns. This frustration level was an indication of their difficulty in articulating deep knowledge (whether first principles or common sense) to provide further demanded justifications.

vii) Because the trio was required to think thoroughly and type many questions and answers, electronic mail was slow from a real-time feedback standpoint. At times the trio and rookie were asynchronized because of the amount of this detail and delay.

viii) The lack of a "chat" mode in the communication protocol prevented the trio and rookie from elaborating on vague phrases like, "you know what I mean..." and "just because". This would have made the exercise better resemble how people actually communicate.

ix)   Three days of knowledge elicitation time, instead of three hours, may be necessary in each of the phases (initial, in-progress, and overall). Only a fraction of the whole body of construction scheduling knowledge was stimulated in our experiment.

Scheduling Evaluation Exercise Consensus. A consensus of the participants concerning the exercise emerged and can be summarized:

i)   There is a sufficient common body of construction knowledge to which KBES technology can be meaningfully applied. The participants felt this way even if only a reliable, noncomputerized set of scheduling guidelines is developed first.

ii)   The limited information approach was accepted as a feasible and practical method for stimulating and capturing construction scheduling knowledge. The rookie in this exercise asked for advice as if he were very naive and uninformed to prod the experts into being explicit. Though their patience eventually wore thin, we obtained reasonably logical, consistent and thorough explanations to questions.

iii) The scheduling professionals became aware of how difficult it is to articulate the thought process and the concepts they utilize intuitively. For future knowledge elicitation efforts we recommend the knowledge engineer express all his or her knowledge about the problem first. Verification by the experts can be a starting point for further elaborations and permit a running start on the exercise.

iv) Owners and contractors share a large number of objectives, though some do conflict. However, a single construction scheduling criticism KBES may be used by both parties. At the least, contractors could anticipate owner concerns by reviewing such a system and prepare their schedules accordingly.

Knowledge base development has largely been characterized to-date by application of ad hoc elicitation techniques. Uncertainty with KBES technology in general and knowledge elicitation techniques in particular have led to confusion surrounding which technique or techniques should be used for a given set of conditions. To lend some clarity to this clouded situation, Appendix I is offered for guidance. It is based on the work of this project as well as efforts by several other investigators (De La Garza and Ibbs 1988; Freiling et al. 1985; Hart 1985; Hoffman 1987; Mittal and Dym 1985; Prerau 87; Trimble et al. 1986; and Trimble 1987).

In this appendix the completeness of a knowledge base is defined at one of four levels. A first-pass is probably inadequate for most serious applications. If used for training purposes, the second-pass category may be sufficient. Diagnostic and classification systems probably require a higher level of completeness, a third- or fourth-pass knowledge base. With the help of Appendix I, we hope other researchers in the construction scheduling domain may be able to more expeditiously and accurately develop their own knowledge bases. In general, the methods chosen in any knowledge elicitation method play an important role in the integrity of the knowledge captured.

5. Acknowledgements

W.E. O'Neil Construction Company, a large building contractor in Chicago, Illinois, collaborated on this study by committing one senior project manager and a junior project scheduler to the development of the knowledge base. Mr. E. William East from the US Army Construction Engineering Research Laboratory (USA-CERL) and a field engineer from the US Army Corps of Engineers participated to articulate an owner's view. A senior consultant from Pinnell Engineering, Inc., a company which provides scheduling services to both owners and contractors, also took part in the knowledge elicitation process. Expertise of several faculty members in the Department of Civil Engineering at the University of Illinois at Urbana-Champaign was also included. We wish to acknowledge the participation of these people in this effort. Many of the ideas in this study owe an intellectual debt to Dr. Michael J. O'Connor of USA-CERL. Accordingly, we would like to express our appreciation and thanks to him.

This material is based upon work supported by the National Science Foundation under Grants Nos. MSM-8451561 and MSM-8613298, and by U.S. Army Construction Engineering Research Laboratory under Project No. AT23-AO-048. Any opinions, findings, conclusions, or recommendations expressed in this paper are those of the authors and do not necessarily reflect the views of the sponsors. Sections of this chapter are part of the Journal of Computing in Civil Engineering, Vol. 4, No. 2, April 1990, Paper No. 24532, and were reprinted with permission from the American Society of Civil Engineers.

Appendix I - Knowledge Base Development Strategies

Method 1: Analysis of Public Domain Knowledge

Description: Analysis of recorded knowledge in published texts and technical manuals.

Benefits: Familiarization with the domain's depth, breadth and terminology. Least time-consuming and expensive acquisition technique.

Disadvantages: May be too general to be of specific use, or dated. Amenable for checklist development.

Products: First-pass knowledge base.

Sequence: First thing to do.

Method 2A: Unstructured Interviews

Description: An expert's response to spontaneous questions about facts, heuristics and procedures.

Benefits: Generation of extensive intuitive and relatively unbiased knowledge. Assists in determining system needs from the standpoint of a user.

Disadvantages: Time-consuming and may result in superfluous information. Lack of structure may confuse the expert and the knowledge engineer.

Product: First-pass and maybe a second-pass knowledge base.

Sequence: Likely conducted after a literature search. Unstructured interviews may be by-passed altogether in favor of more structured meetings.

Method 2B: Structured Interviews

Description: Systematic questioning of an expert on specific procedures and short-cuts. Pre-planned question list, which may be developed after using either Method 1 or 2A, is followed closely.

Benefits: Addition to the knowledge base of less intuitive and more specific domain information. Allows more focus on the problem domain than most other methods.

Disadvantages: Generated information may be biased by the questions asked. Experts may become uncomfortable when unable to articulate less intuitive knowledge.

Product: Second-pass knowledge base.

Sequence: Generally, most effective if conducted after one of the previous methods.

Method 3A: Observation of Familiar Tasks

Description: Expert is studied by watching him/her perform typical, routine tasks freely.

Benefits: Expert is not encumbered with an interview process.   More natural actions and responses are seen. Flexible.

Disadvantages: May be time-consuming because the expert is not concentrating on the problem domain full-time. Little opportunity to get timely explanations from the expert why certain actions were taken.

Product: Second-, maybe a third-pass knowledge base.

Sequence: May be used as an initial review of the problem domain, or may productively follow any of the other methods.

Method 3B: Observation of Tasks with Limited Information

Description: Expert behavior is tracked while typical tasks are performed.   Commonly available information is withheld to see how the expert actually develops assumptions, etc. to complete the information context.

Benefits: Can be applied to stimulate knowledge in very specific domains. Can also be used to fill gaps left by other approaches.

Disadvantages: The limitation of available information may create an unrealistic and artificial setting.

Product: Second or third pass knowledge base.

Sequence: Usually a second or third step, after any of the above methods.

Method 3C: Observation of Tasks with Constrained Processing

Description: Similar to Method 3A except the expert is limited to a set amount of time for responding to force prioritization of procedures, rules and heuristics.

Benefits: Good at stimulating knowledge in a limited domain or when other methods have left gaps in the knowledge base.

Disadvantages: May disorient experts and force hurried choices.

Products: Second or third pass knowledge base.

Sequence: Normally a second or third step in eliciting knowledge.

Method 3D: Observation of Tough Tasks

Description: Study of expert behavior in unusual, unfamiliar or difficult tasks.

Benefits: Helps to focus the domain. Allows a contextual review of expert behavior in highly difficult circumstances. If properly done, may enrich the knowledge base substantially.

Disadvantages: May frustrate the expert. May lead to a knowledge base filled with severe conflicts.

Products: Third or fourth stage knowledge base.

Sequence: Frequently a "mop-up" pass to collect rules, etc. on scattered subjects.

Method 4: Induction

Description: Analysis of attribute-value pairs from a set of case studies, with the purpose of forming governing rules.

Benefits: Addition of structure and new cause-effect relations to the knowledge base.

Disadvantages: Sequence of rule antecedents generated by induction algorithms may not replicate the expert's thought process. Requires an extreme care and insight.

Products: Second-, third-, or even fourth-step knowledge base.

Sequence: Almost always follows an interview process (2A or 2B).

Appendix II - References

Ashley, D. and R.E. Levitt, "Expert Systems in Construction: Work in Progress" *ASCE Journal of Computing in Civil Engineering*, Vol. 1, No. 4, October 1987. pp. 303-311.

Avots, I., "Application of Expert Systems Concepts to Schedule Control," *Project Management Journal,* Vol. 16, No. 1, March, 1985, pp. 51-55.

Clough, R. H., *Construction Contracting,* 4th Edition, 1981, John Wiley & Sons.

Davis, E. W., "CPM Use in Top 400 Construction Firms," *Journal of Construction Management*, ASCE, Vol. 100, No. CO1, March, 1974, pp. 39-49.

De La Garza, J. M., and Ibbs, C. W., Jr., "A Knowledge Engineering Approach to the Analysis and Evaluation of Schedules for Mid-Rise Construction," *Construction Research Series, Technical Report No. 23*, Ph.D. thesis submitted to the Department of Civil Engineering, University of Illinois at Urbana-Champaign, July, 1988.

Feigenbaum, E. A., and McCorduck, P., *The Fifth Generation: Artificial Intelligence and Japan's Challenge to the World,* Addison-Wesley, Reading, MA, 1983.

Freiling, M., Alexander, J., Messick, S., Rehfuss, S., and Shulman, S., "Starting a Knowledge Engineering Project: A Step-by-Step Approach," *The Artificial Intelligence Magazine,* Vol. 6, No. 3, 1985, pp. 150-164.

Gray, C., "Intelligent Construction Time and Cost Analysis," *Journal of Construction Management and Economics,* Vol. 4, 1986, pp. 135-150.

Hart, A., "Knowledge Elicitation: Issues and Methods," *Computer-Aided Design*, Vol. 17, No. 9, November, 1985, pp. 455-462.

Hoffman, R. R., "The Problem of Extracting the Knowledge of Experts from the Perspective of Experimental Psychology," *The Artificial Intelligence Magazine,* Vol. 8, No. 2, 1987, pp. 53-67.

Kneale, D., "How Coopers & Lybrand Put Expertise into its Computers," *The Wall Street Journal,* Vol. 33, November 14, 1986.

Levitt, R. E., and Kunz, J. C., "Using Knowledge of Construction and Project Management for Automated Scheduling Updating," *Project Management Quarterly*, Vol.16, No.5, December, 1985, pp. 57-76.

Michie, D., "Machine Learning and Knowledge Acquisition," in *Expert Systems - Automating Knowledge Acquisition - AI Masters Handbook*, Edited by Donald Michie and Ivan Bratko, Addison-Wesley Publishers, 1986.

Mittal, S., and Dym, C. L., "Knowledge Acquisition from Multiple Experts," *The Artificial Intelligence Magazine*, Vol. 6, No. 2, 1985, pp. 32-36.

O'Connor, M. J., Colwell, G. E., and Reynolds, R. D., "MX Resident Engineer Networking Guide," *Technical Report P-126*, U.S. Army Corps of Engineers Construction Engineering Research Laboratory, Champaign, IL, April 1982.

Ponce de Leon, G., "Schedule Submittals: To Approve or Not to Approve," *Strategem*, Vol. 2, No. 1, Fall, 1984.

Prerau, D. S., "Knowledge Acquisition in the Development of a Large Expert System," *The Artificial Intelligence Magazine*, Vol. 8, No. 2, 1987, pp. 43-51.

Teicholz, P., *CIFE Symposium Proceedings*, Stanford University, March 1989.

Trimble, G., Bryman, A., and Cullen, J., "Knowledge Acquisition for Expert Systems in Construction," *Proceedings of the 10th Triennial Congress of the International Council for Building Research, Studies and Documentation*, CIB86 Advancing Building Technology, Vol. 2, September, 1986, Washington, D.C., pp. 770-777.

Trimble, G., "Knowledge Acquisition for Expert Systems in Construction," *Expert Systems in Construction and Structural Engineering,* Edited by Hojjat Adeli, Chapman and Hall Publishing Co., 1987.

CHAPTER 5

A Diagnostic Aid for
Wastewater Treatment Plants

Catherine D. Perman
and Leonard Ortolano

1. Introduction

SLUDGECADET is an expert system prototype that diagnoses operating problems at wastewater treatment facilities. Its principal elements include a treatment plant model, heuristics used by expert plant operators, and theoretical knowledge of wastewater treatment processes. The intended user of the system, a novice plant operator, supplies the system with information concerning specific operating problems. General information about the plant is stored permanently by the system. At the end of an interactive session, the expert system displays its diagnosis, a recommended remedial action, associated explanations and detailed information about the facility.

The prototype system is built around the heuristic knowledge of five experts in operating wastewater treatment plants. Although these experts use environmental engineering principles in running plants, they are distinguished from novices by their intuitive understanding, based on years of experience, of how plants fail and what it takes to remedy operating problems. [As reported by Zuboff (1989, p. 53), this reliance on intuition is common in operating oil refineries and other "continuous-process production environments."] The goal of the knowledge acquisition exercise was to make explicit the intuitions of the expert operators to create a qualitative model of the process used by experts in diagnosing wastewater treatment plant operating problems and proposing remedies for those problems.

The prototype expert system implements a model of the diagnostic techniques of the five experts. Three of the five experts contributed "generic knowledge," knowledge that is transferable among treatment plants of a similar type. The other two experts provided only specific information relevant to a test case wastewater treatment facility. The diagnostic model was developed during a set of knowledge acquisition activities that included 45 hours of formal interviews and six days of observing the experts conducting their work. The knowledge acquisition was embedded into the standard rapid prototyping paradigm for building expert systems (see, e.g., Malin and Lance 1985, and Hayes-Roth et al. 1983). Findings from the knowledge acquisition effort provide insights into the process of working with multiple experts.

2. Background

There were several factors that led to the identification of wastewater treatment plant operations as a suitable domain for building an expert system. One factor concerned the limited applicability of formal mathematical models. Although models of unit operations exist and they can be implemented using conventional algorithm-based computer programs, the models are typically not suited for field use (see, e.g., Gall and Patry 1989). Another factor leading to the selection of the wastewater treatment plant domain is the difficulty that the United States faces in keeping wastewater treatment plants running to meet design performance specifications and to satisfy requirements of the national effluent discharge permit system (U.S. Government Accounting Office, 1983). Among the reasons for these difficulties is the inexperience of operators; novice operators lack the intuition and judgment that experts have. If these intuitions could be articulated and made explicit as heuristics, there would be good prospects for improving overall plant performance. Expert systems are often well suited for problems in which qualitative, heuristic reasoning is an important feature.

Having identified the domain of diagnosing wastewater treatment plant problems as suitable for developing an expert system, the next step was to narrow the scope of the prototype expert system to a manageable size. One way this was accomplished was by limiting attention to the class of treatment plants that use trickling filters for secondary wastewater treatment and anaerobic digesters for sludge treatment. This restriction had an additional virtue. Because the U.S. Army has many plants using trickling filters and anaerobic digesters, its Construction Engineering Research Laboratory (CERL) became interested in supporting the development of an expert system. CERL provided financial resources, access to experts and a wastewater treatment plant that could be used for test purposes. The U.S. Army wastewater treatment facility at Fort Lewis, Washington, was used as a test site for SLUDGECADET.

The experts participating in building SLUDGECADET included staff at the Fort Lewis wastewater treatment plant and engineers from the ES2 Environmental Services consulting firm of Berkeley, California. The latter group played a key role in defining the scope of the prototype system. A major concern was that the prototype remain small and still be able to represent general problems. At the start of the knowledge acquisition process, the experts emphasized that the selection of appropriate operating problems should be associated with specific unit processes of a wastewater treatment plant. For purposes of defining the scope of the prototype, the typical plant (Figure 1) was categorized as follows: headworks, primary clarification, trickling filters, secondary clarification, chlorination and sludge treatment (i.e., anaerobic digestion).

The experts felt that the feasibility of building a complete system could be established if the prototype could handle a set of problems and unit processes that collectively represented the full range of issues encountered in running a wastewater treatment plant. This meant that unit processes included in the prototype should involve both physical-chemical and biological treatment and that the treatment should involve both wastewater and sludge. It was also considered important to include processes at different stages of a typical plant (e.g., primary clarification near the front end and chlorination near the end).

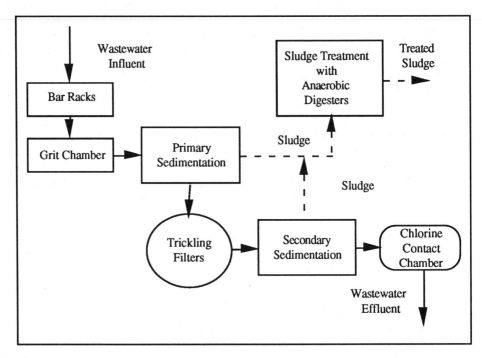

Figure 1. Typical schematic for the type of wastewater treatment plant modeled in the SLUDGECADET system.

After pondering the types of operating problems to be included, the experts concluded that feasibility of building a complete system could be fully examined if all the following issues were treated: malfunctions in physical-chemical and biological processes, mechanical problems, interactions among unit processes, data analysis and interpretation, and external or environmental factors (e.g., changes in the community served by a plant).

Following extensive discussion with the experts, the following three "problem modules" were selected for building the prototype.

The Primary Clarifier Suspended Solids Module. The concern here is with identifying problems associated with malfunctions in primary clarifiers, units in which solids settle out to form a sludge that is collected and pumped to a sludge treatment process. A common symptom of problems in a primary clarifier is a high concentration of suspended solids in the clarifier's effluent. This module represents physical-chemical processes that are inherent to the operation of a primary clarifier and mechanical failures, for example, breakdowns in the sludge pumping system for the primary clarification unit process.

The Anaerobic Digester Supernatant Module. Part of the anaerobic digestion process involves gravity settling of the solid particles in the sludge; i.e., most solids settle to the bottom of the digester leaving a top layer of liquid, called "supernatant", that is relatively clear of solids. Supernatant is frequently returned to the front end of a plant and added to the incoming wastewater. If digester supernatant is of very poor quality, it can significantly change the quality of the incoming wastewater and disrupt the functioning of various unit processes in the plant. The anaerobic digester supernatant module requires consideration of: biological processes in an anaerobic digester; the interaction of failures in one unit process with others in the plant; the behavior of the anaerobic digester in reaction to changes in the local climate (an environmental factor) or service community (an external factor) that could adversely affect plant performance; and the analysis of data (e.g., volatile suspended solids concentrations) used in diagnosing problems in the anaerobic digester.

The Chlorination Fecal Coliforms Module. The effectiveness of chlorination is often measured by concentrations of fecal coliform bacteria, organisms that indicate the presence of contamination by feces of mammals. Problems with chlorination units are often indicated when the concentration of fecal coliform bacteria in a treatment plant's effluent is abnormally high. The chlorination fecal coliforms module examines the following types of problems: physical-chemical process breakdowns that can occur in chlorination, and mechanical problems associated with failures in the chlorination chemical feed control system. This module also involves data analysis to identify trends in wastewater quality parameters (e.g., pH or temperature) used in diagnosing problems with the chlorination unit process.

Collectively, these three problem modules address all the concerns raised by the participating experts regarding issues a prototype must treat to demonstrate the feasibility of building a complete system.

3. Design of the Knowledge Acquisition Exercise

The SLUDGECADET knowledge acquisition exercise emphasized obtaining expertise from one or more individuals and translating that expertise into a testable prototype system. The design of the overall structure of the knowledge acquisition exercise was based on certain objectives and constraints. One objective was to select experts who have had extensive practical experience in operating wastewater treatment plants and would be involved in troubleshooting plant operating problems during the period of the knowledge acquisition exercise. A basic constraint on the scope of the prototype was that there would be only one knowledge engineer available for conducting all tasks associated with knowledge acquisition and implementation. Travel to the expert's offices and to the test site would be limited. It was also assumed that the experts would not have access or time to use software on their own for either knowledge acquisition or prototype development.

The knowledge engineer (the senior author) had academic and professional experience in environmental engineering. Her background, which included treatment plant design and compliance with environmental regulations, did not qualify her as an expert or even a novice in operating wastewater treatment plants. However, she was familiar with the basic terminology, the underlying principles and the engineering literature. She could easily access published information about wastewater treatment plant operations and contribute this information to the knowledge acquisition process.

The SLUDGECADET knowledge acquisition exercise originally envisioned one expert and one test site. Three candidate experts, associated users and treatment facilities were considered. The Facilities Engineer Support Agency (FESA) played a key role in identifying the candidate experts and sites. FESA, a component of the U.S. Army Corps of Engineers, was responsible for assisting the "Corps" staff in enhancing the efficiency of operations of wastewater treatment plants on U.S. Army installations.

One facility considered as a candidate test site was the Fort Lewis wastewater treatment plant near Seattle, Washington. The sanitation chief for the Fort Lewis plant was the expert associated with this plant. A second candidate facility was the Sharpe Army Depot near Stockton, California. In this case, the associated experts were three engineers from ES2 Environmental Services (hereafter referred to as the ES2 engineers), Berkeley, California. The ES2 engineers were working as consultants to FESA in an effort to improve operations at the Sharpe Army Depot. The third candidate facility was the Fort Ord wastewater treatment plant near Monterey, California. In this instance, the experts were engineers from V.J. Viccione Engineering of Virginia, another consulting firm working with FESA. The original plan was to select one test site and one expert from among those available at the two consulting firms and the various candidate wastewater treatment plants.

What are the appropriate criteria for selecting an expert? It could be argued that experts used in building an expert system should be the "best in the field." In this context, it is difficult to define "best." Moreover, even given a suitable definition, it is an unworkable ideal to expect the best expert to be available and to have other attributes that are important in building an expert system. As a practical matter, an expert is someone who is regarded by those interested in solving the problem as having an outstanding record of accomplishment.

For this research, a candidate expert was defined as a person who met the following criteria:

• Diagnosis and repair expertise: Although the individual should have an understanding of the design and theoretical aspects of wastewater treatment engineering, her or his practical experience and acknowledged expertise should be with the daily operations of a plant.

• Interest in participating: The expert should be willing to participate in developing the expert system and enthusiastic about the research project.

• Familiarity with Army facilities: The expert must be familiar with the kinds of wastewater treatment facilities that are used at U.S. Army bases (e.g., facilities with trickling filters and anaerobic digesters).

• Ability to communicate: The expert must be articulate and willing to analyze and share her or his problem solving methods.

• Computer literacy: The expert should be comfortable with computers. This is not a general requirement; however, for this research, it was felt that a basic level of ability with computers would make the knowledge acquisition process easier. It was postulated that an expert familiar with computers would be enthusiastic about and patient with the process of designing and building an expert system.

• Compatibility with the knowledge engineer: The expert and the knowledge engineer should find it easy to work with each other.

• Time availability: The expert should have enough time available for knowledge acquisition, and should be located reasonably close to the knowledge engineer's work place (Stanford University).

The process of selecting the expert and the test site from the candidates involved several interviews and site visits. The knowledge engineer went to: (1) the Sharpe Army Depot in Stockton, California with a staff engineer of ES2 engineers; (2) Fort Ord in Monterey, California with a staff engineer of V.J. Viccione Engineers; and (3) Fort Lewis in Seattle, Washington to meet with the sanitation chief.

The original concept of selecting a single expert was not followed. Five experts were selected after visiting each site, discussing the planned expert system with the experts, observing them at work and learning about their working methods. The knowledge engineer designated three of the ES2 senior engineers as experts. Fort Lewis was chosen as the test site. The Fort Lewis test site offered two additional experts: the sanitation chief and the operations foreman. Both have many years of experience in successfully operating the Fort Lewis treatment facility and were very enthusiastic about working on the research project. Each expert met all the previously noted criteria and, in addition, had some other attributes. The ES2 engineers were interested in learning about expert systems technology. They had already created a computer program to compare treatment plant design performance to actual performance and were interested in learning about new computer programming techniques. They also had personal computers in their offices that could be used for demonstrating the implemented versions of the expert system. The Fort Lewis sanitation chief is a computer enthusiast; in addition, he has personal computers at his office and was planning on purchasing a personal computer for the wastewater treatment plant.

Because the final selection included more experts than originally anticipated, it was necessary to modify the original design of the knowledge acquisition exercise to accommodate characteristics of the problem solving methods of all participating experts. Because the ES2 engineers only knew about the Fort Lewis plant from design, operation and maintenance manuals, not personal experience, they offered general expertise in diagnosing problems and recommending remedial action. They could supply expertise applicable to the class of plants of which the Fort Lewis facility is a member. On the other hand, the sanitation chief and the operations supervisor offered site specific expertise about the Fort Lewis facility. This expertise would be used to implement an application of the expert system for the test site. Thus, the ES2 engineers would supply information applicable to all trickling filters plants with anaerobic digesters, and the Fort Lewis experts would supply information for a specific site. By separating the knowledge sources into general and site specific, the knowledge acquisition exercise contributed to an expert system design objective: transferability among wastewater treatment plants.

The approach of separating general knowledge from site specific information has foundations in an expert system paradigm called model-based reasoning. The model-based reasoning paradigm uses a mixture of data structures (called frames or objects) and rules to implement an expert system. This approach allows the knowledge engineer to write rules about a system or process that is modeled with data structures. One feature of model-based reasoning is that data structures can be used to create general, transferable models that can be easily modified to describe a specific facility or site. The model-based reasoning used in SLUDGECADET is discussed in detail by Perman (1989), and earlier examples of model-based reasoning are described by Koton (1985) and Nardi and Simons (1986). The SLUDGECADET knowledge acquisition exercise reflected the model-based reasoning paradigm; the experts explicitly directed the knowledge engineer in distinguishing between general and site specific knowledge about wastewater treatment plants with trickling filters and anaerobic digesters.

At the start of the knowledge acquisition exercise, the knowledge engineer and the experts met at a few U.S. Army wastewater treatment facilities, including the Fort Lewis plant. The early phase of knowledge acquisition included watching the experts work, discussing case histories and collecting information that would allow the knowledge engineer to design the scope of the prototype system. During the first few meetings with ES2 Environmental Services and Fort Lewis personnel, the knowledge engineer asked the ES2 engineers and Fort Lewis staff for case histories of wastewater treatment plant problems and how the problems were solved. These discussions produced some examples, but it also led to some problems. At first, the experts had trouble coming up with specific examples and could think of only one or two operating problem case histories. As the discussions continued, the experts were able to think of more case histories. However, the effort involved in structuring and organizing information from these examples was frustrating to both the experts and the knowledge engineer. Examples that the experts mentioned were cases that were readily available to the expert's memory, but they did not necessarily provide a representative or thorough sample of cases that the experts had actually experienced (cf. Tversky and Kahneman, 1974, p. 27). Hours were spent without yielding an organized body of acquired knowledge.

After assessing results from the initial visits with the experts, it became clear to the knowledge engineer that the knowledge acquisition interviews needed to become more productive. Consequently, the knowledge engineer decided to structure subsequent interviews around developing test cases based exclusively on the previously noted problem modules. The modules helped to organize and focus the attention of the knowledge engineer and the experts during the knowledge acquisition exercise. The knowledge engineer and experts at ES2 Environmental Services established a schedule for a series of knowledge acquisition meetings. It was agreed that the best way to begin would be to choose and thoroughly discuss one test problem for a single unit process. The ES2 engineers recommended starting with the primary sedimentation tank because it had an appropriate level of complexity for the first problem module.

In response to questions from the experts about expert systems technology, the knowledge engineer also presented a brief seminar to the ES2 engineers. It included an introduction to expert systems and the knowledge acquisition process and a demonstration of a simple rule-based system written in INSIGHT for diagnosing problems in a trickling filter. [INSIGHT is a rule-based shell produced by Level Five (1985).] Interestingly, the experts at ES2 had done some reading on expert systems; one expert had obtained a "freeware" inference engine and had experimented with rule-based representation. During the seminar, the experts asked many questions about the structure of the expert system and spontaneously started sharing their ideas and advice on the structure of both the proposed system and the knowledge acquisition process. The seminar provided a forum for the experts and the knowledge engineer to discuss the expert system development process and afforded the experts a chance to think and comment about the process as a whole.

4. Summary of the Knowledge Acquisition Exercise

Knowledge acquisition began with site visits with the ES2 experts and Fort Lewis staff. The visits were used to delineate the scope of the prototype. Then, information for the prototype was elicited during formal interviews between the experts and the knowledge engineer. The various activities that constitute the knowledge acquisition exercise are summarized in Table 1. It characterizes the sessions with the experts and provides a general history of the entire exercise. The early sessions, which involved watching the experts work, informal discussions, site visits, etc., provided valuable information for designing the prototype architecture and functionality. The details of the prototype took form during interviews with the experts about each problem module; the methodology used in these interviews is described in detail below.

The transfer of knowledge was accomplished using several methods, all based on direct and personal interaction between the knowledge engineer and one or more experts. For example, informal conversations established a basis for working relationships and offered opportunities to define basic terminology. As another example, formal seminars, given by the knowledge engineer, provided more detailed information to the experts and potential users about artificial intelligence, expert systems and the objectives of the planned expert system.

The process of transferring knowledge from the experts to the computer program was carried out in cycles. Each cycle consisted of the following steps: interview, transform, implement, demonstrate, review and refine. Interviews were used to gather new knowledge that was then transformed into programmable form. Once the knowledge was implemented as a component of the prototype expert system, that component was demonstrated to the experts. Their comments were used to refine the implementation. Each cycle began by collecting the more general expertise from the ES2 engineers and ended by having the Fort Lewis experts add site specific expertise.

During each interview, the knowledge engineer and an expert worked on a specific problem module. Elements of the expert's problem solving techniques were revealed by examining how the expert discovered a problem symptom, how the causes of various symptoms were diagnosed, and how a particular remedial action was recommended. The rules articulated in the context of the problem module came from the expert's own experiences. Rules were constructed from an expert's reactions to: (1) rules developed by the other participating experts, and (2) from reactions to problem solving heuristics found in the literature. After each of these interviews, the knowledge engineer formalized the knowledge and added it to the developing expert system.

The next step in transferring knowledge consisted of getting the experts' reactions to the implemented problem modules by providing a demonstration of the prototype expert system. The experts' reactions to the demonstration were valuable in several ways. The most important was in testing the correctness of the work done by the knowledge engineer in transferring knowledge from the expert to the program.

During a demonstration, the knowledge engineer and an expert could verify that the newly implemented portions of the prototype system were emulating the expert's problem solving techniques. The demonstrations also provided a forum for reviewing the clarity and ease of use of the interface associated with the new portions of the system. This knowledge acquisition cycle was repeated as each problem module was added to the SLUDGECADET system.

Phase	Sessions	Time	Location	Format
Preliminary Sessions	3	6 days	Sharpe Army Depot, Fort Ord, Fort Lewis	Site Visit
	1	2 hours	ES2 offices	Seminar, Scoping Session
Primary Clarifier Module	1	2 hours	ES2 offices	Interview
	1	2 hours	ES2 offices	Demo, Review
	1	6 hours	Fort Lewis	Seminar, Demo, Review
Anaerobic Digester Module	2	2 hours	ES2 offices	Interview
	4	6 hours	ES2 offices	Review
	2	3 hours	ES2 and Fort Lewis	Demo, Review
Chlorination Module	1	2 hours	ES2 offices	Interview
	3	6 hours	ES2 offices	Review
	1	3 hours	Fort Lewis[a]	Demo, Review
Validation Pre-Test	4	4 hours	Stanford	Interview, Questionnaire
Internal Expert	1	3 hours	Berkeley	Interview, Questionnaire
External Expert	2	3 hours[b]	Stanford	Self-study Questionnaire

[a]An ES2 expert was also present at Fort Lewis
[b]This includes estimates for time that the internal expert spent working alone.

Table 1. Summary of Knowledge Acquisition Activities.

5. Incorporating Validation into Knowledge Acquisition

An integral part of the knowledge acquisition process was checking if the experts agreed with both the wording and content of the recorded information. This constitutes an element of expert system validation: information destined to be encoded in the system is tested with an "internal" expert (i.e., an individual contributing information to the expert system). The expert's review and approval was incorporated into the knowledge acquisition process in several ways. The knowledge engineer formatted the interviews so that the expert could periodically check and revise information recorded during an interview. Sometimes, testing was as simple as the knowledge engineer repeating the expert's words or rephrasing statements to check that the knowledge engineer understood the content as well as the implications of the expert's statement. Repetition alone was a powerful tool. During the SLUDGECADET interviews, when the expert heard the knowledge engineer repeat a phrase or statement, the expert often rethought it and changed the wording or elaborated the concepts.

Entire meetings were allocated to checking and revising results of previous interviews. For example, some sessions were devoted completely to reviewing diagnostic flow charts and giving the expert an opportunity to make revisions. These charts were originally drawn by the expert, then redrawn (and annotated with comments tape recorded during previous sessions) by the knowledge engineer. While redrawing these charts, the knowledge engineer would often notice apparent contradictions or missing information in the recorded information. Redrawing the charts raised numerous questions that might have otherwise gone unasked. During the work for both the anaerobic digester and chlorination problem modules, revisions were significant enough for both the experts and the knowledge engineer to realize that the information in initial diagnostic flow charts would have provided a poor foundation for an expert system. Additional review sessions were scheduled until both the expert and the knowledge engineer were satisfied that flow charts were correct and complete.

Results from the second knowledge acquisition session for the chlorination fecal coliforms module provide an example of how a diagnostic flow chart was changed during a revision. The example concerns a portion of the flow chart that represents the first few parameters that the experts checked before proceeding with more detailed data gathering. Figure 2 shows this portion of the flow chart at the beginning of the session, and Figure 3 shows the revised version of the same material. Both figures display a flow chart that starts at the top and works down from left to right. A revision was made to the section of the flow chart inside the dashed boxes. In Figure 3, the experts added a test for flowrate to the left branch in the box and added a check of the chlorine feedrate to the right branch inside the box. This revision was one of many changes made during this knowledge acquisition session.

Demonstrations of the SLUDGECADET expert system provided still other opportunities for the experts to evaluate the accuracy of the encoded knowledge. While using the system or watching a system demonstration conducted by the knowledge engineer, the experts commented on the appearance and the wording of the diagnostic knowledge. In addition, they commented on whether the system was understandable and appropriate for the intended users.

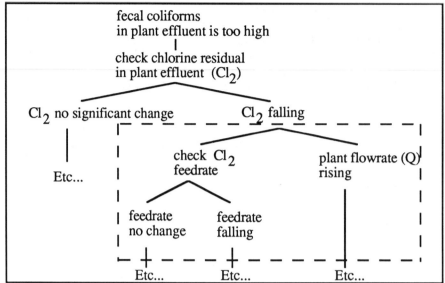

Figure 2. Diagnostic flow chart at the start of the second knowledge acquisition session for the chlorination fecal coliforms problem module.

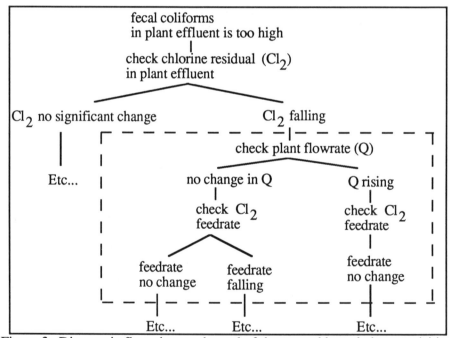

Figure 3. Diagnostic flow chart at the end of the second knowledge acquisition session for the chlorination fecal coliforms problem module.

6. Final Results and Validation

Results from the knowledge acquisition activity took the form of a general model with two parts: a model of treatment plants (implemented using the framed-based knowledge representation and object-oriented programming features of KEE) and a model of the diagnostic process, which was based on the rules obtained during knowledge acquisition. For more on the knowledge representation used in SLUDGECADET, see Perman (1989). To apply these general results to a specific facility, such as the Fort Lewis wastewater treatment plant, it is necessary to create appropriate instances of the objects in the general model. For example, the Fort Lewis plant has four primary clarifiers. To create a model of the Fort Lewis plant it is necessary to create four instances of the primary clarifier object in the general model. Each such instance contains specific details regarding size, shape, design flowrate, etc. needed to describe the corresponding primary clarifier at the Fort Lewis plant.

The validation of the SLUDGECADET application for the Fort Lewis plant required investigating the acceptability of the diagnoses and recommendations it produces. During knowledge acquisition interviews, the experts and knowledge engineer worked together to minimize discrepancies before any implementation occurred. In addition, the experts' reactions to the prototype demonstrations were useful in verifying that newly implemented portions of the prototype emulated their problem solving techniques. The program demonstrations allowed the experts to examine the system's results and explanations. Sometimes the internal experts examined detailed reports of the results and reasoning of the expert system. These reports allowed the results of the program to be critiqued without the distraction of running the system. These activities did not constitute a complete verification of the prototype because the demonstrations were limited in scope. Nevertheless, this style of knowledge acquisition attempted to make knowledge base validation an integral part of system development.

A second and more formal validation exercise took the form of validation experiments conducted after the three problem modules of the prototype were completed. This involved three test cases based on actual operating problems experienced at the Fort Lewis plant during the summer of 1988. The experiments involved preparation of test case "dossiers" containing information describing the Fort Lewis plant and data on the particular test case operating problem (e.g., gas bubbles are rising to the surface of primary clarifier number 3). Using only the information in these dossiers, independent diagnoses were produced by three agents: two experts and SLUDGECADET. One expert was internal. The second expert was neither involved in knowledge acquisition nor familiar with SLUDGECADET (i.e., he was an "external expert").

A comparative analysis was performed using results produced by SLUDGECADET and each of the two experts. Each of these results was also compared with an empirical benchmark, namely, the problem diagnosis and remedy produced by experts operating the Fort Lewis plant at the time the problem symptoms actually occurred. The validation experiment results were encouraging and provide a basis for further refining the prototype. Details on the validation study are given by Perman and Ortolano (1989).

7.   Analysis of the Knowledge Acquisition Methodology: Conducting Interviews

The basic format of the knowledge acquisition exercise was a personal interview during which the knowledge engineer elicited information from an expert using questions and comments. The knowledge engineer conducted the interviews using an informal conversational style; the structure of the problem modules was used to guide the interview content and objectives.

The characteristics of the SLUDGECADET knowledge acquisition interviews varied with the goals and needs of the session. The earliest interviews, during which the knowledge engineer and experts had informal discussions and the knowledge engineer observed the experts working, were unstructured (Mishler, 1986). The discussion flowed from the topic at hand and the knowledge engineer did not work from a set list of questions. At that point, the knowledge engineer simply didn't know all the necessary questions and which questions were important. Part of each unstructured interview involved the knowledge engineer learning about the subject matter and, in addition, about how to interview the individual expert. The unstructured interview created an environment for experts to talk introspectively and in great detail about their work methods.

During these early interviews, the experts made it clear to the knowledge engineer that they were interested in more than answering a series of questions. They wanted to ask questions about artificial intelligence, about expert systems, about specific software and hardware, and they wanted to contribute to the design of the knowledge acquisition process and the SLUDGECADET expert system. For example, the ES2 engineers had general and specific recommendations regarding the scope and focus of the knowledge acquisition exercise. They requested that the sessions be taped, and they spontaneously took over the activity of drawing diagnostic flow charts during each interview. If the knowledge engineer had adhered to a predetermined script for the interview, a great deal of information would have been lost.

Later knowledge acquisition interviews were more formal. They were based on sets of questions defined by the subject of each problem module. For example, questions for the primary clarifier suspended solids module focused on how to diagnose the problem of abnormally high suspended solids in the primary clarifier effluent. When the knowledge engineer and the expert were developing the diagnostic flow chart for a problem module, the interview was unstructured and used a directed format (Mishler, 1986). During an unstructured and directed interview, the knowledge engineer guided the discussion using the subject of the problem module and a list of topics or general questions. One aspect of the unstructured interview format that the knowledge engineer found critical was the freedom to develop unanticipated questions in response to the expert's answers. In addition, the expert was encouraged to be creative, to contribute questions and to digress or "tell stories." Digressions were an important part of the unstructured, directed interviews.

In each interview, a substantial amount of time was spent clarifying terms used and statements made by experts and the knowledge engineer. The knowledge engineer frequently misunderstood the experts or found their comments unclear. A conversational interview format allowed the knowledge engineer to repeat questions and get needed explanations from the experts. Similarly, the experts had their own difficulties in understanding the knowledge engineer. The most common offense in terms of confusing the experts, was the knowledge engineer's use of artificial intelligence or computer jargon. Defining jargon or avoiding it altogether eliminated much confusion.

When the knowledge engineer and the expert were reviewing a new version of a diagnostic flow chart, the interview became more structured. These interviews used the existing version of the flow chart as a guide in determining the order and content of the questions asked by the knowledge engineer.

During the interviews, the expert was encouraged to take as active a role as possible. For example, since the knowledge acquisition process required the preparation of drawings and diagrams, the expert actually drew those diagrams thereby directly recording his exact wording. This minimized the role of the knowledge engineer as secretary. If the knowledge engineer tried to write accurate notes, then she wouldn't have had the time to listen carefully and formulate relevant questions. With the expert involved in recording problem solving techniques, the knowledge engineer was free to listen and ask questions. Although interviews were taped to assure that details would be captured, making verbatim transcripts was considered overly time consuming and unnecessary for this research. By working with the diagnostic flow charts and by having the experts actively create the charts and revisions, the need for transcripts was reduced.

The structure of the knowledge acquisition process helped in obtaining accurate representations of the expert's diagnostic process. Experts drew flow charts and, in subsequent interviews, had opportunities to review and revise those charts and previously articulated rules. The experts could not be expected to describe their reasoning and problem solving procedures by responding quickly to a straightforward set of questions. Learning about their reasoning processes was a subtle task that required the cooperation, introspection and review of the expert as well as the understanding, listening and support of the interviewer.

8.   Analysis of the Knowledge Acquisition Methodology: Using Interview Roles

The SLUDGECADET knowledge acquisition interviews were a type of conversation that was "a joint product of what the interviewees and interviewers [talked] about together and how they [talked] with each other" (Mishler, 1986, preface). How the expert and the knowledge engineer talked to each other was as important to the success of the knowledge acquisition exercise as the content. [These observations have been made in other contexts by Forsythe and Buchanan (1988) and Werner and Schoepfle (1987).]

As with any conversation, the knowledge acquisition interview was an emotional event as well as an intellectual event, and the emotional dimensions of the interview presented difficulties to both the knowledge engineer and expert. For example, the knowledge engineer sometimes had trouble admitting that she knew less than the expert in the area of the problem domain. At the start of the knowledge acquisition exercise, the knowledge engineer found herself in an awkward position. She had a graduate degree in environmental engineering but did not have any practical experience in wastewater treatment plant operations. The graduate training gave the knowledge engineer a sufficient level of domain knowledge to understand the technical vocabulary of the experts, but it did not provide much insight into how experts diagnose problems. As the experts explained their diagnoses, she thought "I probably should know this material, but I don't. Will I appear stupid and poorly educated if I ask about this? And will the experts lose all their confidence in me if I admit that I don't understand what they are saying?"

From the experts' point of view, the combination of the knowledge engineer's high level of education and low level of practical experience made it difficult for them to judge how much of their problem solving approach to explain. The experts initially had certain expectations as to what a Ph. D. candidate should know about their domain. In later discussion, they described their initial feelings in terms like these: "I have a Ph.D. candidate here, I'd better not make a mistake in front of her," and "I don't need to explain this; she must already know all about it." In this context, the experts felt awkward because they were placed in the position of teaching and explaining aspects of the domain that might seem trivial or elementary. The knowledge engineer let the experts know that even the most basic material may not be obvious to her and might need elaboration.

By assuming specific roles, the experts and the knowledge engineer alleviated these difficulties. During the SLUDGECADET knowledge acquisition exercise, each expert assumed the role of "mentor" during the interviews and the knowledge engineer assumed the role of "apprentice" while discussing the diagnostic process. These roles eased the initial discomfort felt by the experts and the knowledge engineer, allowed the knowledge engineer to ask "stupid" questions and encouraged the experts to elaborate on their problem solving techniques in great detail.

Clearly, the knowledge engineer's high level of domain knowledge influenced the knowledge acquisition exercise in both positive and negative ways. On the positive side, the knowledge engineer was familiar with domain terminology and theory. This familiarity enabled her to understand the experts' explanations with little additional effort. A knowledge engineer with less domain knowledge would have needed to spend additional time learning the basic terminology and theory of the domain. On the negative side, the knowledge engineer had to contend with some embarrassment and awkwardness during knowledge acquisition interviews because of her level of domain knowledge. She had to explain the limits of her knowledge to the experts and make it clear when she needed additional information. A knowledge engineer with less domain knowledge would have had an easier time defining the limits of her knowledge. Another negative outcome concerned an instance in which the knowledge engineer held a professional judgement different from that of the expert. She made a point of arguing for her position and that produced a very awkward moment. Because this instance occurred late in the knowledge acquisition exercise, there were no serious consequences. If it had occurred in one of the first few interviews, the knowledge engineer's ability to work with that expert might have been very adversely affected.

9. Analysis of the Knowledge Acquisition Methodology: Maintaining the Expert's Motivation

Although the experts started the process as willing and enthusiastic participants, over the course of knowledge acquisition, social and organizational factors diverted the experts' attention from the expert system building process. For example, more than once during the knowledge acquisition process with the ES2 engineers, a participating expert had to postpone a scheduled interview or shorten an interview to satisfy other obligations. These ranged from a personal leave for an expert who took two weeks off due to the birth of his child, to an extended consulting trip, to an unusually heavy work load that required ending an interview session earlier than planned.

Given the many sources of distraction and the lengthy and methodical character of the knowledge acquisition process, the knowledge engineer had to make efforts to maintain the experts' interest and willingness to participate. There were several factors that seemed to keep the experts' interest level high. The experts were gratified by talking about their expertise and experiences. Some experts found being cast in the role of "expert" flattering, and it gave them a sense of responsibility for ensuring that the system accurately represented their expertise. Some experts found that the intellectual challenge of helping to develop an expert system was itself a form of compensation. Others found the exercise to be a learning opportunity. For instance, the ES2 engineers were curious about the expert system technology and expressed great interest in it. The introductory seminars that the knowledge engineer presented were, in part, directed toward satisfying their curiosity with the hope that explaining a new technology to the experts would enhance their commitment to the development of SLUDGECADET.

The ES2 engineers seemed keenly interested in the progress of the knowledge acquisition process. They frequently asked questions about their own performance as: "Is this [information] what you [the knowledge engineer] are after?" "How am I doing?" and "Are we making progress?" Other questions dealt with less immediate issues, such as the design of the expert system or the academic progress of the knowledge engineer. Responding to these questions helped maintain the experts' good will and enthusiasm.

System demonstrations provided another source of motivation. The experts were always interested in seeing revised versions of SLUDGECADET in operation. This provided tangible evidence that their ideas were being put into action. The demonstrations also helped the experts visualize the ultimate product of the expert system building exercise.

A non-rigorous, but nonetheless interesting indication that the experts were compensated for their efforts in ways noted above is that they chose not to accept financial payment for their work. Although remuneration was offered to the ES2 engineers, none was accepted.

10. Analysis of the Knowledge Acquisition Methodology: Working with Multiple Experts

As detailed previously, the experts were split into two groups. Each group contributed different types of knowledge to the SLUDGECADET expert system: the ES2 engineers provided the "generic" knowledge, which was transferable among similar wastewater treatment facilities, and the Fort Lewis experts provided site specific information. A basic assumption of the SLUDGECADET expert system architecture was that these two types of knowledge, while complementary, could be kept distinct.

Each group of experts, associated with generic or site specific knowledge, influenced the type of information receiving emphasis during the interviews. For example, certain topics would be passed over by the ES2 engineers because they felt the knowledge was site specific. One advantage of this split is that it clearly defined the limits of each group's subject matter and reduced the amount of information covered by each group. Since the experts were contributing time that would normally have gone to other professional responsibilities, any technique that reduced the time required for their participation and sharpened the focus of the interviews made the process more efficient and less burdensome.

Using multiple experts turned out to be constructive rather than problematic, and it enhanced the SLUDGECADET knowledge acquisition process (cf., Mittal and Dym, 1985). The ES2 engineers supported and elaborated on each others work. For example, they actively collaborated during the initial interviews concerning the "anaerobic digester problem," with one assisting another when the first had reached an impasse. In other sessions covering the same problem, the experts worked together to revise and correct each other's results. During demonstrations of SLUDGECADET, all the ES2 engineers watched and offered opinions that were actively discussed until agreement had been reached.

Working with more than one expert was practical. Since multiple experts were available, the knowledge engineer could schedule frequent sessions without demanding too much time from any one expert. When unanticipated scheduling conflicts arose, the experts could trade or split time commitments. In this way, the knowledge engineer could proceed with knowledge acquisition at a reasonable pace without depending heavily on the availability of a specific person.

11.   Conclusions

Conclusions from the knowledge acquisition exercise used in building SLUDGECADET included insights into the nature of expertise in operating treatment plants. Various manuals and guides for treatment plant operators provide information on the heuristics used by expert operators. However, the results from the knowledge acquisition activities provided heuristics that had not previously been elaborated. These concern the ways in which experts order their searches for problem causes and the ways in which they make decisions regarding calls for additional diagnostic test data. The knowledge acquisition activities also highlighted the importance that expert operators place on identifying how problems in one unit process can cause difficulties in downstream processes. Heuristics concerning interactions among treatment processes are not generally highlighted in manuals on plant operations.

The knowledge acquisition activities also provided insights into working with experts. For both the knowledge engineer and the experts, assuming specific roles greatly eased tensions that had been felt at the beginning of the process, and it made the interviews more productive and pleasant. Having the experts play an active part of the record keeping also greatly reduced the amount of notes that the knowledge engineer had to take. In addition, the act of drawing diagnostic flow charts seemed to keep the experts alert and more aware of what they were saying. The use of multiple experts was another successful feature of the knowledge acquisition process. The ES2 engineers continually helped each other contribute to the knowledge acquisition process. Having the ES2 experts provide transferable knowledge and Fort Lewis experts give site specific knowledge illustrates how separate groups can successfully contribute different types of knowledge to an expert system.

The SLUDGECADET knowledge acquisition exercise demonstrates the value of putting special effort into the selection of experts. Much preliminary work was done on identifying criteria for selecting experts and these criteria were applied carefully. The resulting choices were excellent in that the experts were articulate, enthusiastic and knowledgeable. Experts with such attributes are likely to be extremely busy and that certainly was the case for the experts who helped build SLUDGECADET. Indeed, although they had agreed to participate in this project, there was always the sense that they needed to rush off and get back to their "real work." In an ideal world, the experts would be given leave from at least a portion of their regular responsibilities, so that they could devote themselves more fully to the expert system building process.

The participating experts from both ES2 and Fort Lewis made the SLUDGECADET knowledge acquisition exercise successful. Without their cooperation and their willingness to share intuitive diagnostic skill that was heretofore tacit, the expert system prototype could not have been built.

Acknowledgements

We thank the experts who participated in developing the SLUDGECADET prototype for their patience and thoughtful contributions. They are Dave Sullivan, Roy Monier and Drew McIntyre of ES2 Environmental Services, Berkeley, California and Dave Hanky and Dick Pitzen of the Fort Lewis Wastewater Treatment Facility, Fort Lewis, Washington. Research support was provided by the U.S. Army Corps of Engineers Construction Engineering Research Laboratories, Champaign, Illinois under contract number DACA88-86-D-0008. IntelliCorp Inc., Mountainview, California, generously provided the use of their hardware and software during the development and testing phases of this research. We also thank Anne Steinemann and Kristy Kocher, both of the department of Civil Engineering, Stanford University, for reviewing an early draft of this chapter.

REFERENCES

Forsythe, D., and Buchanan, B.G. (1988). "An Empirical Study of Knowledge Elicitation: Some Pitfalls and Suggestions." In Lehner, P.E. and Adelman, L. A(eds.) *Methods in Knowledge Engineering*, special issue of IEEE Transactions on Systems, Man and Cybernetics.

Gall, R.A.B., and G.G. Patry. (1989). "Knowledge-Based Systems for the Diagnosis of an Activated Sludge Plant" in Party, G.G. and D. Chapman (eds.) *Dynamic Modeling and Expert Systems in Wastewater Engineering*. Lewis Publishers, Inc., Chelsea, Michigan.

Hayes-Roth, F., Waterman, D.A., and Lenat, D.B.(eds.) (1983). *Building Expert Systems*. Addison-Wesley, Reading, Mass.

INSIGHT. (1985). Computer Software. Level Five. Melbourne Beach, Fla.

Knowledge Engineering Environment 3.1 (KEE 3.1). (1988) Computer Software. IntelliCorp. Mountain View, Ca.

Koton, P.A. (1985). "Empirical and Model-Based Reasoning in Expert Systems." *Proceedings of the Ninth International Joint Conference on Artificial Intelligence.* August18-23, Los Angeles, 1,297-299.

Malin, J.T., and Lance, Jr., N. (1985). "Feasibility of Expert Systems to Enhance Space Station Subsystem Controllers." *Proceedings of SPIE Conference: Space Station Automation.* September 17-18, Society of Photo-Optical Instrumentation Engineers, Bellingham, Washington. 580.

Mishler, E.G. (1986). *Research Interviewing: Context and Narrative.* Harvard University Press, Cambridge, Mass.

Mittal, S., and Dym, C.L. (1985). "Knowledge Acquisition from Multiple Experts." *AI Magazine.* 6(2), 32-36.

Nardi, B.A., and Simons, R.K. (1986). "Model-Based Reasoning and AI Problem Solving" *Workshop on High Level Tools for Knowledge Based Systems.* Sponsored by AAAI, OLAIR and DARPA. October 6-8, Columbus, Ohio.

Perman, C. D. (1989). *Improving the Performance of Wastewater Treatment Plants: an Expert Systems Approach.* Ph.D. Dissertation. Department of Civil Engineering, Stanford University, Stanford, Ca.

Perman, C. D., and Ortolano, L. (1989). "Testing a Prototype Expert System for Diagnosing Wastewater Treatment Plant Operating Problems." presented at *The First International Conference on Expert Systems in Environmental Planning and Engineering.* September 21-22, Cambridge, Mass.

Tversky, A., and Kahneman, D. (1974). "Judgement under Uncertainty: Heuristics and Bias." *Science.* 185,1124-1131.

U.S. Government Accounting Office. (1983). *Wastewater Treatment Dischargers are not Complying with EPA Pollution Control Permits.* GAO/RCED-84-53. U.S. Government Printing Office, Washington, D.C.

Werner, O. and Schoepfle, G.M. (1987). *Systematic Fieldwork Vol. 2: Ethnographic Analysis and Data Management.* Sage Publications, Newbury Park, N.J..

Zuboff, S. (1989). *In the Age of the Smart Machine.* Basic Books, Inc., New York, N.Y.

CHAPTER 6

Knowledge Acquisition for an Expert System
for Handling Customer Inquiries on Water Quality

Richard M. Males
Judith A. Coyle
Walter M. Grayman
Robert M. Clark
Harry J. Borchers
Beth G. Hertz

1. Introduction

An expert system dealing with customer inquiries for a water utility has been developed as part of an ongoing cooperative agreement between the North Penn Water Authority (NPWA) and the US Environmental Protection Agency. The system is designed to allow non-technical, administrative personnel to handle routine inquiries about water quality that were normally handled by laboratory personnel. The system is based primarily on the expertise of a single individual, Judy Coyle, formerly water quality manager at NPWA. (Males, 1991)

The system has been evaluated in prototype by NPWA personnel, and gives good agreement with 'expert' conclusions. As of this writing, the system is being re-worked slightly in preparation for implementation by non-technical personnel. Knowledge acquisition for the system was conducted in a single, six-hour session in January of 1989, and a prototype system was provided to NPWA in April of 1989. Total costs for this phase of work are estimated to be under $20,000.

It should be noted that this expert system was initially developed primarily to familiarize the project team with expert system approaches and technology, but later became of interest to NPWA for fielding as a functional system.

105

2. Background

The North Penn Water Authority, Lansdale, Pennsylvania, serves approximately 18,000 residential, commercial, and industrial customers in the Montgomery County area of Pennsylvania, north of Philadelphia. The water system consists of a main system, and some disconnected satellite systems. At the time of this study, water was supplied from some 50 wells extending into a fractured/ fissured rock aquifer, and from treated river water, purchased from the American Water Works Service Company (formerly the Keystone Water Company). Another surface water source has since been developed. Depending upon the time of year and demand, water supplied to a particular location may be river water, well water, or a mixture of the two. The well water has a high degree of hardness, and the river water has seasonal odor problems resulting from algal blooms. The well water is chlorinated, but not fluoridated. Many of the homes in the area use water softeners.

NPWA has been working, since 1988, with the Drinking Water Research Division, US Environmental Protection Agency, Cincinnati, Ohio, in a cooperative research agreement to examine the use of expert systems in operation of water distribution systems to improve water quality. The original purpose of this project was to explore the combination of local operational expertise with simulation models of hydraulics and quality for the water system. As an initial step in this research, a sample problem relating to water quality was selected, to gain experience with expert systems tools and approaches, prior to embarking on further studies involving more complex problems of system operation. This paper deals exclusively with the initially selected sample problem, described below.

3. Problem Selection

An initial literature review and contacts with individuals experienced in the field of expert systems, suggested that the scope of the initially selected problem be limited and clearly defined:

> "Experience has shown ... that the most successful projects are those with fairly modest objectives, taking about six man-months to develop. The problem should be clearly defined and bounded, i.e. it should be easy to describe what the expert system should do, its use and scope. Furthermore, the chosen problem must be sufficiently important to warrant investigation, while being technically feasible." (Hart, 1986)

A meeting was held at NPWA to discuss appropriate problems that would be significant, of interest to NPWA personnel, and sufficiently complex and with embodied spatial and water quality characteristics to be valuable for the research effort. After considerable discussion, the problem of handling the authority's 'customer complaints' (later expanded to 'customer inquiries') about water problems was identified by NPWA personnel and selected as the appropriate problem for the prototype expert system. Among the features of the problem that suggested that it would be a good candidate for expert system development are:

a) It is a real problem, of significance to the utility, as opposed to a fabricated one - in fact, it was proposed by NPWA personnel, as an alternative to an initially suggested problem relating to determining the sources of constituents found in water samples, as being more realistic, and more important;

b) An expert was available - Judy Coyle, Water Quality Manager at NPWA, and formerly chemist and laboratory director at NPWA;

c) Expertise and familiarity with the specifics of the NPWA system are required to handle the calls - in fact, when the individual [Judy Coyle] who normally handled the calls went on vacation, a significant amount of time was spent in training others to take the customer calls;

d) The expertise was departing - Judy Coyle was switching jobs within the NPWA, and would not be available to handle the calls;

e) The problem is reasonable in size - there are limits to the questions that are asked, and the responses that are given, and much of the information can be categorized (both inquiries and responses). The problem is specific, and is not open-ended.

f) The problem is interesting. NPWA personnel were enthusiastic about working on the expert system development.

g) The result, while specific to NPWA, would have generic aspects common to other water utilities handling customer inquiries.

Once the problem to be addressed was selected, a general description of the customer inquiry process was prepared, as follows:

Customers call NPWA frequently. A customer query usually is in the form of a telephone call. [No examination of other inquiries, e.g. letters, was contemplated]. Calls usually arrive at a central administrative desk. The administrative desk will attempt to answer the question if a specific problem is known (i.e. there has been a fire in an area, resulting in low pressure or brown water), or refer it to the appropriate department (e.g., construction inquiries, etc.). Judy Coyle will get the water quality problems (e.g. odor in water, brown water, do I need a water softener?), and, occasionally, other queries come to her inadvertently - these, she refers to the proper department.

There may be as many as five or six calls a day, more at times of specific problems. All calls are logged into a notebook. Depending upon the particular question, there is a typical set of response actions on the part of the utility, including: sending an individual out to sample at the customer's site; recording the information but not sending someone out; sending a form letter containing answers to frequently asked questions; answering the question directly; etc. A variety of criteria are used to determine the response. The individual handling the call asks general questions about the customer location, character of problem, and frequency of problem, and, depending upon the problem, refines the question in an attempt to determine: a) what the problem is, and b) what to do about it. It is not always possible to determine the cause of the problem over the phone. These cases dictate a 'what to do about it' response involving taking one or more samples, to further identify the problem.

The purpose of the expert system was, broadly, to assist untrained individuals (perhaps front office personnel, or less experienced laboratory personnel) in handling customer inquiries, but beyond this general expectation, no specific goals or functions for the expert system were defined at this time. Later, a more detailed view of the desired functions of the expert system was defined.

4. Knowledge Acquisition Process - Preparatory Steps

In order to characterize the nature of the customer inquiries, prior to embarking on detailed knowledge acquisition activities, tape recordings were made (with the customer's agreement) of conversations between Judy Coyle and the customer. About 45 minutes of conversation were recorded and transcribed for review by the project team members who would participate in the knowledge acquisition process.

A typical portion of the transcript, dealing with a brown water problem, is as follows ("J" is Judy Coyle, "C" is the customer):

J: You live in Harleysville, and you had brown water?
C: Right, since last evening - I just filled our washer up, and it's dark brown.

J: Would you say it's worse this morning than it was yesterday?
C: Yes. It started late in the evening.

J: Have you spoken with your neighbors yet?
C: I haven't checked with them, but I can do that now and call back.

J: Let me ask you a couple of questions in the meantime. Is it also in your toilet. Is it all through your house?

C: Yes, everywhere. There are no clothes in the washer. I took them out and ran it through, I was thinking it might clear out the line as it filled. The other spigots I had running for a while ten or fifteen minutes, still coming through brown.

J: OK, as far as the magnitude of the brown, how would it look compared to say a cup of tea?

C: Dark, darker.

J: OK, strong tea we're talking about here. Do you have any kind of a water softener or anything?

C: No we don't.

J: I'm not right at the moment aware of anything that would cause a problem out there, so I'm going to have to do some checking around to see what I can find out.

A second portion shows additional types of questions, relating to an odor/taste problem:

J: Do you have any kind of a water softener?

C: Yes.

J: Is all of your water softened? Like the water your husband was washing the car with - could that have gone through the softener too?

C: No. I'm not sure if that was hard or soft. The water I got out was hard, we drink the hard water, and I have a hard water tap. I'm not sure if our outside spigot is hard or soft, I don't think it is softened, I think it's just hard.

J: It sounds like it was definitely not the problem. Sometimes the water softeners give an odor to the water, but that doesn't sound like it if you got it from your tap.

C: No it was our hard water. We always drink the hard water.

J: It couldn't be ice cubes, because your husband smelled it and ...

C: We didn't have ice cubes in the water, we just got it right from the tap.

The value of this effort cannot be overstated. It provided the knowledge acquisition team with a good feeling for the nature of the questions that were asked, and the categories of problems that showed up. It clearly indicated some of the repetitive questions that were asked by laboratory staff, and showed much of the 'human' nature of the interaction - some callers were clearly irritated and needed to be placated, others were interested in 'chatting' about peripheral subjects.

While the tapes were being made, Judy Coyle made an attempt to categorize the types of problems that she encountered from the customer inquiries (e.g. brown water, taste/odor problems, inquiries relating to hardness, etc.). In addition, she prepared a list of standard information that she would acquire from the customer during the phone call.

During these initial steps, an effort was made to avoid forcing a structure onto the problem - rather, the interest was in simply revealing what actually takes place in the expert-customer interaction, and providing some initial categorization to allow for focusing the detailed knowledge acquisition.

5. Detailed Knowledge Acquisition

A meeting was held at NPWA on January 13, 1989 to perform the knowledge acquisition. Judy Coyle and Harry Borchers (Executive Director) of NPWA provided the 'domain expertise', i.e. the knowledge about the problem. Other NPWA staff members were invited to participate at various times during the meeting, to provide specific information as appropriate. At the time of the knowledge acquisition, Judy Coyle was transferring jobs within NPWA, and responsibility for handling customer queries on water quality was being shifted to a new individual. Thus, the expert system was seen as a way of transferring Judy's expertise to the new individual.

The 'knowledge engineers' were the project consultants, Walter Grayman, Richard Males, and Mark Houck. Grayman and Males had previously worked with NPWA, and were familiar with the area, the operation, and the staff of NPWA, but had not done major work in expert systems. Houck was not familiar with the area, but had reviewed the transcripts of the customer calls, and literature describing the NPWA, and had previous experience in knowledge acquisition, expert system development, and in the general field of water resources.

The knowledge acquisition meeting was tape-recorded, and some 6 hours of tapes were transcribed (in computer-readable format as an ASCII data file) providing a complete record of the discussions.

The general structure of the session was free-form in nature. Initially, Houck provided an overview and some discussion of general concepts and experience with expert systems. This led to 'knowledge elicitation' on the topic, under Houck's guidance, starting with determination of the goals that exist (on the part of NPWA) in handling each customer inquiry, and following on with categorization of problems and responses. Houck strongly emphasized that the objective of the day's effort was to find out how Judy Coyle processed and responded to the information, rather than attempting to fit that into a pre-conceived framework or fit a particular computer program. The methodology used for knowledge elicitation was open-ended, with no specific formal structure. The transcripts of customer query phone calls provided examples of methods that Coyle used to handle queries in a variety of situations. Coyle had also roughly categorized the nature of the queries, providing a solid basis for exploring her responses to each one. In addition, Coyle had designed a data base for recording customer calls, again providing definition of the facts that she felt should be obtained during a phone call. Roughly the first half of the discussion was general talk 'around the topic', getting the dimensions of the problem and a structure for examining it put together. The second half of the discussion was more focused on the handling of the specific categories of inquiries, attempting to identify the specific questions that are asked, the language used to describe problems (i.e. how do people describe a chlorine odor?), and the questions and logic she goes through in processing these problems. A newsprint pad was used to record the category by category results.

During the course of discussion, the group successively defined:

a) the general facts about the problem that are determined at the start of each phone call;

b) the set of responses that exist for a customer query;

c) the specific phrases or descriptions that lead to categorizing the query into a particular type, together with the associated follow-up questions and appropriate responses for that particular type of query; No attempt was made, at this stage (or later in the project), to quantify any responses or certainty levels (i.e. "I believe this to be a brown water problem with a certainty of .5"). In addition, no specific expert system shell or knowledge representation format was in mind during the knowledge acquisition.

An example of a portion of the knowledge acquisition session, dealing with a specific category of problem, is as follows ("JC" is Judy Coyle, "RM" is Richard Males, "WG" is Walter Grayman):

JC: Do you want to start with an easy one? How about milky water, white milky bubbly water? The first thing that tells me is that it is well water.

RM: How is this described?

JC: My water is cloudy, they will say it is murky white. Sometimes they have even said it is dirty, but when I ask questions, I find out it is white. If anything that they tell me makes me think that it is this bubbly water problem from Harleysville, which is what I am leading up to, their address clues me in, when they start to describe it, I will tell them about this phenomenon that we have, and ask them, when they take a glass of water, does it look white like milk, if they let it sit, does it clear from the bottom or from the top, because that means it is bubbles clearing, and not something settling out. Because I want to make sure that what I am telling them is relating to what they are experiencing.

WG: And this is only in Harleysville?

JC: Harleysville, but it does occur in other areas, well areas. Usually from Harleysville. If someone called me from Souderton, I wouldn't rule it out.

RM: Do you check odor at this time?

JC: I usually ask is there any odor associated with it. Quite commonly, there is a chlorine odor associated with it, because the bubbles are driving the chlorine out.

WG: Do they understand how chlorine smells?

JC: I don't ever say 'does it smell like chlorine'? I say "does it have an odor, and ask them to describe it, if they can, and if they can't then I'll give some choices.

RM: What choices do you usually give?

JC: Again, like something rotting, chemical, like a swimming pool.

6. Refinement of Goals for Expert System

As noted above, the subject of the expert system had previously been loosely identified as a method of responding to customer telephone complaints relating to water quality. During the knowledge acquisition discussions at NPWA, this was expanded to handling general customer queries regarding water quality (i.e. some calls that come in are not specifically complaints, rather are requests for general information). Over the course of the day, further clarity as to the nature and purpose of the system developed. The ultimate intent was to have a system that would:

a)   record customer queries into a data base;
b)   advise the user of the system as to an appropriate response to the customer query;
c)   where appropriate, print out the necessary documents to accompany the response (e.g. work orders, information letters to the customer, etc.);
d)   record the actual and system-advised responses in a data base;
e)   provide capability to record the follow-ups to individual queries, i.e. if a sample was taken, what were the results.

Thus, the system would serve as a combination data-base and advisor, and provide, over time, a tracking system for handling customer queries, and, to some degree, automate the process of generating paperwork associated with customer queries.

7. Computerized Management of Transcript

The transcript of the knowledge acquisition session was some 67 pages in length. Due to the nature of the discussions, a particular topic might be addressed in a number of places within the transcript, or a concept might involve a number of categories. In an attempt to simplify the organization of the information obtained from the knowledge acquisition, the computer-readable transcript was imported into the Lotus Agenda™ software package. Agenda is a personal information manager software package that is designed to assist in cross-referencing information that is in text format. Agenda was selected because it was available, not necessarily as the best software package of its kind for this purpose. Through the use of Agenda, it was possible to cross-reference paragraphs in the transcript under a variety of different topics. Thus, any reference within the transcript to the topic 'color' could be brought up, or references that combined 'color' and 'odor'. Inasmuch as Agenda allows for creation of synonyms, references to 'odor' could also be referred to in the actual discussion as 'smell', 'color' could be 'blue', 'brown', etc. As new categories or topics were deemed to be important, or new synonyms discovered, they could readily be added within the Agenda structure.

A portion of the category structure used to examine the transcript within Agenda is shown below. Categorization allows for a number of options. Items separated by a ',' must exist simultaneously. Thus, in the first line, the specification is for any paragraph in which the word odor appears, but not the word musty or the word earthy. Under "bubbles", any one of the words "bubbles", "milky", "spurting", or "murky" will assign the paragraph to the bubbles category. Words enclosed in parenthesis are to be treated as a single word, and words preceded by ! are negated. As noted above, it is simple to revise this category structure as new insights into how the information should be organized develop.

Sample Portion of Agenda Category Structure for NPWA Transcript
  Problems
    Odor,!musty,!earthy
    Taste
    Color
    Bubbles;milky;spurting;murky
    Bacteria;BACT;coliform
    Chlorine;(swimming pool);chlorox;clorox
    Pressure
    Rusty;brown;rust
    Giardia
    musty;earthy;dirty;geosmin
    lead
  Location
    Harleysville
    Souderton
    Hatfield
    Towamencin
    Lansdale
    Worcester
  Attitude
    hostile;angry;crazy;anger;hostility
  Actions
    sample;bottles
    information

Although it is claimed that software such as Agenda can be useful to organize undifferentiated information into appropriate groups or categories, no such attempt was made in the current project. The retrieval capabilities, however, were found to be useful while developing the codification of the knowledge base and creating the actual expert system, as a means of checking back to the transcript about particular topics. The Agenda software is noticeably complex, with a long learning curve, and the full capabilities could not be explored within the current project. It seems clear, however, that such techniques can provide valuable benefits in categorizing and retrieving the large amount of text information that becomes available when a knowledge acquisition session is transcribed in computer-readable format.

Later on in the project, a brief experiment was made in which the computerized transcript was analyzed by a custom-written C language program that essentially counted the words contributed by different individuals. The intent of this effort was to examine the contributions over time of different individuals during the knowledge acquisition (in part to determine whether or not the knowledge engineers were doing most of the talking), and tabular and graphical displays showing this information were prepared. The experiment proved interesting, and, given the existence of computer-readable transcripts, such automated analysis might be valuable in the future to aid in research on the knowledge acquisition process itself, for example to determine the amount of 'prompting' that is done by knowledge engineers, through examination of the average length of responses by the expert.

8. Codification and Organization of the Knowledge Base

The knowledge acquisition efforts resulted in a variety of information about the customer query process. The typical handling of a call was defined, specific information that is acquired for most calls was identified, and standard responses were described.

Typical Call Handling

Once a call is received, a set of general information is obtained. The customer usually first identifies the problem without prompting (e.g. 'my water smells bad'), and Coyle proceeds to get basic information (address, name, phone number, etc.). If she recognizes that the customer is not in fact an NPWA customer (based on location), then she will refer the individual to the appropriate water utility. She also notes the tone and attitude of the caller (hostile, confused, etc.). After obtaining general information and information about the problem, she will attempt to confirm or clarify the caller's description of the problem. For repeat calls relating to the same/similar problem, the questioning may be abbreviated. During the course of a conversation, she will frequently provide general information (e.g. a description of hardness as a typical problem, a description of the wells in the area) as well as specific information about the problem in question. She then chooses a course of action as a response - either simple provision of information to the customer, obtaining a sample, or looking elsewhere for additional information.

General information obtained for each caller includes: customer name; service address; name of subdivision, if applicable; home/daytime phone number. Other questions typically asked initially are: Has this problem been seen before?; When did the problem start?; Do the neighbors have the problem?; Is the water in the house softened?; Just the hot water, or both hot and cold water?; Is there any other type of home treatment device?

In addition, Coyle implicitly evaluates the 'hostility level' of the caller, as well as whether or not the caller has a legitimate question [a very small percentage of callers feel that 'people are trying to poison their water', or have other odd reasons for calling]. It is considered advisable to inform sampling personnel of the attitude of the person making the inquiry, if that may be a problem.

Standard Actions and Problem Categories

Based on the knowledge acquisition, a set of standard 'responses' of NPWA to a customer query, that cover almost all situations, were defined:

a) Obtain a sample at the customer site.
b) Pick up a customer-collected sample.
c) Send an informational letter.
d) Check with the Engineering Department.
e) Drop off sample bottles
f) Give information to the customer directly
g) Measure chlorine residual at the customer site
h) Transfer the query, in-house
i) Send NPWA employee to area to 'look around'
j) Send laundry rust remover
k) Ask customer to do something
l) Immediately visit the customer

Coyle has also defined a number of standard 'query categories' that organize most of the quality-related calls:

a) White, Milky, Bubbly Water
b) Questions (not problems or complaints)
c) Brown Water
d) Hardness
e) Chlorine Taste and Odor
f) Other Odor
g) Color Other Than Brown
h) Bitter Metallic Taste and Odor (possible copper)
i) Low Pressure
j) Musty/Earth Odor/Taste
k) Giardia

These standard responses and query categories formed the basis for the structuring of the knowledge base.

PseudoCode Approach

For each category, a description of the category, together with the questions used to confirm the category, and resulting actions, was prepared. In preparing these, a method called 'pseudocoding', borrowed from computer science, was used. Pseudocode is an intermediate form between english language and computer language, used to describe a problem in a manner that can then be readily translated into a particular computer language. Pseudocoding is frequently used as an alternative to the diagrammatic 'flowcharting' methodologies of computer programming. An example of pseudocode for dealing with a chlorine odor problem is as follows:

IF chlorine odor:
THEN:
    tell customer that wells will be checked [usually satisfies them]
    ask engineering to check on wells

IF customer upset:
THEN:
    do chlorine residual test on-site
    advise them to store water in loosely capped jar in refrigerator

    This format is relatively easy to extract from the discussions (and in fact was used to record information on the newsprint pad during the knowledge acquisition session), and can be incorporated into a computer implementation fairly simply. An example of a category description for bubbly water is as follows:

Category: White, Milky, Bubbly Water
Descriptors: Water is cloudy, milky white, dirty.
Location: Harleysville
Confirming questions:
    Does it look like milk?
    Does a glass of water clear from the bottom to the top?
    Is the area near Harleysville?
    Is there an odor? [usually chlorine, driven out by bubbles]
    What kind of odor? (rotting? like a swimming pool? chemical? musty? earthy?)
    Is the water running normally, or is it spurting?
Actions:
IF	non-spurting, milky white, clears from bottom to top:
THEN:	describe problem (typical Harleysville water), give information (no health problem)
IF	spurting:
THEN:	call engineering (problem is air, entrained by pumps)
IF	musty/earthy odor:
THEN:	sample in winter, or if obviously not river water [musty/earthy odor typical of river water when there are algal blooms - spring / summer]
IF	odor other than chlorine or musty/earthy:
THEN:	sample

Here, the 'then' portion of the category description leads to one or more of the standard responses defined previously.

In constructing these category descriptions, frequent reference was made to the transcription of the knowledge acquisition session, and to the data organized within the Agenda software, described above. The ability to refer back to the transcript through computer-assisted means was important, as frequently the notes taken during the knowledge acquisition session were incomplete, or missed later or earlier comments that revised the manner in which a particular problem is handled.

9. Selection of Appropriate Expert System Framework and Technology

It is considered advisable to avoid selecting a particular expert system approach until the structure of the problem is clear, which should come after the knowledge acquisition phase. This avoids 'forcing' the knowledge structure into what is supported by a given tool. The most common type of expert system (at least for beginners) is a rule-based system, in which knowledge is represented largely as a set of 'if A, then B' rules. At the end of the knowledge acquisition session, it was clear that a rule-based approach to expert system design would be a reasonable fit to the actual manner in which Judy Coyle processes information.

A variety of low cost rule-based systems and system shells are available for personal computers (the IBM compatible PC had previously been selected as the hardware environment for the project). The ability to access a database from the expert system was desired (in order to get information about a customer's account, or to examine previous customer inquiries). Thus, the primary criteria for selection of an expert system tool were:
- a)   handling of rule-based problems;
- b)   data base access available;
- c)   IBM-compatible computer environment.

Inasmuch as the problem was clearly rule-based, the use of an expert system shell to rapidly develop and test the logic and rule structure seemed advisable. Once the logic and rules were so defined, the possibility of using an alternative representation (e.g. a programming language) to provide better user interfaces, could be explored. A low-cost expert system shell (VP-Expert, Paperback Software, Berkeley, CA) was selected, because of its low cost, wide use for simple expert systems, and also because of its advertised data base access features, and a variety of methods it provides for user input. The intent was to concentrate on the logic at first, and leave the niceties of the user interface to later in the project.

10. Development, Testing and Status of the System

After familiarization with the VP-Expert package, and experimentation with a variety of methods for developing the rules and defining the goals (i.e. particular variables to be searched for by the expert system), the complete system was developed as a prototype. The purpose of the initial prototype system was to test the logic of the rule base, and to demonstrate the system features. Once the structure and tools were in place, development went fairly rapidly, based on the pseudocode description of each problem. Each category was developed and tested individually, and then the entire system was tested as a whole by the developers. It must be noted that testing by developers is not the same as testing by the eventual users. The testing by developers is designed to insure that the logic matches the logic that has been codified from the knowledge acquisition (not that it is the 'correct' logic). Testing by the user is to insure that the system works in a real world situation, and that the user interface and logic are appropriate.

Forty-four rules were required for the system. The initial system lacks some of the features of the desired design, in particular the capability to look up a customer in a database. This was due to a bug in the VP-Expert system's methods of handling external database information. Other features were implemented in only a rudimentary fashion, e.g. the printing of work orders and form letters, in order to demonstrate the feasibility and method of implementation.

The system, as developed, used a fairly simple, text-based user interface, with selection of options from menus of choices. VP-Expert also offers a more sophisticated user interface allowing for data entry into forms, and choice of items with a mouse, but this requires a somewhat more sophisticated hardware environment than the basic text-oriented system. A small portion of the system was implemented, as a demonstration, in the more sophisticated, mouse-oriented user interface, in order to allow users to assess the difference and choose a preferred interface. When demonstrated to NPWA personnel, the mouse-oriented approach was clearly favored.

The prototype system was installed and demonstrated at NPWA, with the intent that it be verified in use. After a short test period, in which the logic apparently performed adequately, the system was set aside, as other priorities and personnel changes took place at NPWA. (In fact, the original expert, Judy Coyle, left North Penn Water Authority, and her job was assumed by another individual). Thus, testing was done by individuals who were, in reality, testing a knowledge base derived from an 'expert' no longer at NPWA, providing an excellent test of the logic, as well as a training tool for the new employees as to the nature of water quality concerns in the NPWA system.

Testing was done in conjunction with handling of actual customer inquiries. Rather than using the system in 'real time' while dealing with customers on the phone, notes were taken during the phone calls, and the system was then tested against these notes at a later time. The system performed adequately on most of the problems, with only minor logic flaws relating to the handling of odor complaints. An additional category of problem, sickness or skin irritation, was requested, and a number of additional 'usability' features were requested (the ability to clear a previous entry, the ability to recall data previously entered, linkage with account numbers, etc.). Also desired was the ability to store information about NPWA's response to the problem. Thus, the expert system is clearly being viewed not solely as a consultant/adviser technology, but also as a data base and full-functioned system.

The prototype system has been evaluated by NPWA to determine whether or not to proceed to full-scale implementation, for a system incorporating all of the desired functionality, that can be used by non-technical personnel. The prototype system clearly shows that the embedded logic of the knowledge base is adequate, but additional usability in the form of an improved user interface, and better data base links, are desired. NPWA has determined to go forward with the additional development, with the intent of creating a full-scale system.

11. Conclusions

The prototype expert system demonstrates that a reasonable expert system can be developed by relatively inexperienced expert system users, given certain preconditions. The main conclusions to be drawn from the effort to date are as follows:

1)   Confinement of the problem is essential.

The customer inquiry problem is a bounded, finite problem, that could be examined with reasonable resources. The problem is real, and of interest to the potential users. The literature on expert systems clearly indicates that these are important pre-requisites to the success of any potential expert system, and the current project confirms this.

2)   There is a high need for trust, and a high need for familiarity with the problem, in the knowledge acquisition process.

Knowledge acquisition is usually, in the expert system literature, viewed as something that happens between a 'domain expert', i.e. the individual knowledgeable in the particular area to be addressed, and a 'knowledge engineer', i.e. someone familiar with the processes and techniques of building expert systems, but not necessarily familiar with the domain. In the current project, the individuals serving as knowledge engineers were in fact quite familiar with NPWA, with the problems of water utilities in general, and had worked with NPWA personnel for a number of year. Thus, it was possible to use a lot of technical terminology, and raise a lot of questions, that probably would not be available to someone new to the water field. The individuals fulfilling the role of knowledge engineer could often, by virtue of their knowledge of the water field, and certain situations at NPWA, point to substantive problems or areas that needed to be examined.

In addition, there was a high degree of trust present during the knowledge acquisition process - certain areas, which might have been viewed as sensitive, could be thoroughly explored. The knowledge acquisition session was notably light in tone, with jokes, kidding, and a good deal of interchange amongst the parties.

Most of the information necessary for developing the logic of the expert system was obtained in only a single six-hour session. In retrospect, it simply does not seem possible that knowledge acquisition would have been nearly as fruitful if it had been necessary for the experts to explain the basics of water quality measures, hydraulics, and other technical concepts to a "knowledge engineer". This suggests that it is perhaps better to train someone skilled in the domain (in this case water) in the technology of expert systems, rather than to rely upon a knowledge engineer with no knowledge of the field being examined.

3)   Programming/Systems Analysis skills are clearly needed in the development of an expert system.

In spite of the claims made for expert system shells that they can be operated by unskilled individuals, it was clear from early on that, in order to develop the expert system, an individual experienced in programming would be necessary. As noted above, the 'knowledge base' of VP-Expert, and other expert systems, is, in essence, a programming language. Other expert systems have components that interview an expert and build rules, but, at this point, it seems unlikely that this approach could be very fruitful. Thus, the intermediary of a programmer/analyst (or 'knowledge engineer') familiar with the particular technology selected, seems necessary.

4)   The problem is inherently modular, but there are few capabilities in the form of tools or techniques, that explain how to appropriately structure the development of the expert system for ease of development, use, and modification.

The tool selected, VP-Expert, offers few capabilities, and even less instruction, in how to properly structure an expert system. As noted above, development of the system is, in fact programming. Current thinking on software design emphasizes modularity and structured software design, i.e. the breaking up of problems into small parts, which are then developed, tested, and combined to achieve the total desired functionality. For the current problem, some of the desired modules would be:

    a)   capture of information from the user;
    b)   examination of a data base of client accounts;
    c)   consultation of the rules to define the proposed actions;
    d)   display of the proposed actions, and provision of a simple means to modify them;
    e)   logging of the information to a customer inquiry data base;
    f)   printing of form letters;
    g)   printing of work orders.

Thus, the consultation of the rule base is simply one of the desired modular functions. In a traditional programming environment, the problem could be readily structured to handle all of the items above, in the appropriate sequential order, with the exception of the consultation of the expert system rule base. Within the programming environment of VP-Expert, however, there is no simple way to organize the flow of actions. This is severely limiting to the overall design of the system. The VP-Expert capabilities are excellent for rapid prototyping and design and testing of the rules, but are perhaps insufficient for full implementation. This is again consistent with discussions in the literature relating to the use of expert system shells.

5)   The interest in the system, on the part of the potential users, is not simply in the 'capturing of expertise', but in the totality of the system and its functionality.

It was clear, through the knowledge acquisition and evaluation of the system, that the additional features of the proposed system were of equal importance to the expert system component. In particular, the ability to log customer inquiries into a database, to access a database of the customers, and to automatically print work orders and form letters, were considered very attractive and important features of the system design. It is doubtful that the 'advisor' portion of the system alone, would be of significant interest. It should be noted, however, that once the expert system logic was demonstrated to be reasonable, this part of the effort was taken almost as a 'given', i.e. there was little questioning of the ability of the system to provide reasonable advice consistent with the behavior of the 'expert'. Rather, the focus of questioning, at least at this stage, was on the user interface and data base access issues.

6)   The development of such systems is clearly iterative.

The literature on expert systems development indicates that the process is highly iterative. In the present case, where all parties were generally familiar with the topic, and the 'expert' was capable of articulating the situation almost directly in 'pseudocode', perhaps a few cycles of iteration have been cut from the normal path of development. Nonetheless, it is obviously necessary to go back a number of times to the experts to get clarification.

7)   Computerized management of the transcript is valuable, and computerized analysis has interesting potential.

The ability to rapidly locate any phrase or topic during the knowledge acquisition, through the use of the information manager and word processor, proved to be a valuable feature, greatly assisting in the development of the specifics of the rule base. The computerized analysis of the transcript also proved interesting, and the possibility of extending this concept to more closely examine the knowledge acquisition process itself appears worthwhile.

Appendix. References

Hart, Anna, (1986). *Knowledge Acquisition for Expert Systems*. McGraw Hill, New York, N.Y.

Males, R.M., Coyle, J.A., Borchers, H,, Hertz, B., Grayman, W.M., Clark, R.M., "Expert Systems in Water Utility Management: Handling Customer Inquiries." submitted to *Journal, American Water Works Association*.

CHAPTER 7

Testing An Expert System For the Activated Sludge Process

Wenje Lai
P. M. Berthouex

1. Introduction

In many important ways, wastewater treatment plant operation is an information management problem. Plant performance could be improved if the operator were equipped with convenient and effective methods for using his data. Viewed in this way, a key control problem is how to aid the operator in diagnosing operating conditions and taking appropriate control actions.

This paper briefly describes an expert system (ES) that grew out of work on statistical methods for process control. A management information system, consisting of a data base management program coupled with packages for graphical data analysis, was developed in order to facilitate data analysis. While studying process control strategies, several interesting applications of expert systems were discovered (Vassos 1987, Kramer 1987, Johnston 1985, Huang et al. 1986, Beck et al. 1978, Tong et al. 1980, Maeda 1984 and 1985, Vitasovic and Andrews 1987) and it seemed logical and interesting to incorporate an ES into the management information system as well. Because developing an ES was not the primary goal of the project, our ES rule base is not as elaborate as some others. Still, we believe it can be of interest because of the way that it is integrated with the data base and statistical analysis packages, and because of the approach taken to test it with operators. This paper will focus mainly on testing the accuracy and benefits of the program. The elicitation, calibration, and testing of the rule base at one plant is fully explained.

The expert system to be described is only intended for making routine decisions. It is not expected, for example, to solve problems of plant start-up or of recovery from toxic shock. The goal of this expert system is to free the human expert from routine decision making in order that he/she may be available to tackle the truly difficult situations where creativity, intuition, and experience are vital ingredients in solving the problem at hand. The more specific objectives were (1) to develop methods for interrogating domain experts to collect control rules, (2) to collate and study the advice given by the ES and the actions actually taken by operators in order to improve the ES, (3) to examine the system's accuracy in detecting problems in the treatment process and in providing appropriate advice on control actions, and (4) to investigate the effectiveness of the expert system in improving the performance of less skilled operators. These development and evaluation steps were done using sets of case problems derived from actual operating data at the treatment plants (Lai 1989).

2. The Management Information System

The wastewater treatment plant operator's problem in collecting, storing, manipulating, and interpreting data is an information management problem. STATEX is a statistical and expert management information system developed to help the operator organize and utilize the data collected daily in the treatment plant for improving process control decisions (Berthouex et al. 1989). It combines a database management system, a system for statistical analysis of operating data, and a rule-based expert advisory system. The integration of these systems puts modern information management technology into the hands of any treatment plant operator and provides him a convenient and effective method for using his data in day-to-day treatment process control.

The programs were written in dBASE III and Lotus 1-2-3 and are run on IBM-compatible personal computers. No special computer capabilities are required. Tests have shown that plant operators easily and quickly learn to use the system. No knowledge is required of the programming languages. Even though the program has been tested with treatment plant operators, it is not presently being used routinely at any treatment plant.

The data entry system provides the necessary links to the data analysis system and to the expert advisory system. Through this system, the operator can easily and quickly access current or past data. The data input system will immediately and automatically compute process variables (e.g., detention times, overflow rates, sludge age) and do material balance calculations. (See Appendix II for a glossary of terms.)

The data analysis system will display tables or graphs. In day to day operations, a table of numbers is vastly inferior to graphical displays. The human eye and brain are intuitively skillful at recognizing graphical patterns; therefore, data plotting methods are given emphasis. They are the best tool for identifying abnormal trends or conditions. Various plots of the data are available in the system, including daily data plots, X-Y scatterplots, moving averages (7-day, 30-day, exponentially weighted), cumulative sum (cusum) charts, and external reference distributions (Berthouex and Hunter 1983). A reference distribution might represent daily values, moving averages, differences between successive observations or between two variables, or for computed variables, which may have complex statistical properties, the theoretical statistical properties of which are unknown (Berthouex and Hunter 1983).

3. The Expert System (ES)

Figure 1 shows the structure of the ES. The knowledge base contains the codified expertise extracted from the domain experts. The form given to the knowledge base depends upon the purpose (e.g., planning, diagnosis, prediction, control) of the expert system. It is quite natural to have heuristic control rules such as:

IF MLSS is high THEN increase the sludge wasting rate (WAS).

Since MLSS is measured as a numerical value, the fuzzy term 'high' can be given a numerical definition. The knowledge base will contain rules for translation of fuzzy terms to numerical settings. A typical example might be as follows:

IF MLSS > 2300 mg/L THEN it is high

The inference engine automatically accesses the data base whenever the operator has finished his data entry. This provides timely advice based on the most complete and current information. The inference engine reads current values for the variables in the control rules, compares the recorded values with the state limits specified in the rules, and classifies the variables according to the fuzzy terms used in the rule syntax. The inference engine then checks whether the observed conditions match conditions specified in the IF portion of any control rules. If a match is found, the rule is triggered and the user is presented with advice to take one or more control actions.

The tool box is a special feature of the expert system. It contains utilities to help define the fuzzy terms for the rules and to assess the quality of the advice being offered. The tool box accesses the historical database to display time-series plots of the data, to identify time periods when the operation is under control based on the quality of the effluent data, and to generate external reference distributions. The median, upper, and lower 5% levels of the reference distributions are automatically given. The median represents the desired condition of the variable, based on past conditions. The upper or lower 5% levels indicate extraordinary conditions. These levels might be used as the initial trial definitions for 'high' or 'low' action levels. The user may request other upper and lower percentage levels to be generated.

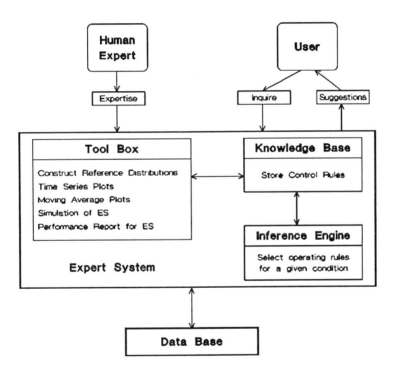

Figure 1. Structure of the Expert System

The tool box offers a simulator to help investigate proposed action levels for the variables of interest for a number of decision test cases representing a variety of different control situations. The simulator applies a set of rules that may be under consideration to sets of sample data that represent a selected range of operating conditions. As these cases are analyzed, the simulator counts the number of correct and incorrect decisions made. Two kinds of decision errors were made by the expert system. A "miss" is when the expert system failed to give a warning or suggest a control action when, in the opinion of the domain expert, there was a recognizable problem and some action was needed. A "false alarm" is an expert system recommendation for a control action when the domain expert thinks none is needed. A performance index was defined as a weighted measure of how many wrong decisions were made and also of the kind of error:

$$P_j = W1\ N1_j + W2\ N2_j$$

where P_j denotes the performance index based on run j, $N1_j$ is the number of misses and $N2_j$ is the number of false alarms observed in run j, W1 is the weight for misses, and W2 is the weight for a false alarm. We require $W1 + W2 = 1.0$. Assigning $W1 > W2$ would indicate that misses are considered a more serious kind of error than false alarms. A lower index value represents consistently more correct control decisions. If there are too many wrong decisions, the action levels of the control rules can be modified to improve the expert system performance.

Occasionally, the operator should generate a performance report that list the control actions actually taken by the operator and the actions suggested by the expert system. Table 1 shows a portion of a performance report that is prepared by the tool box. It can also display a variety of operating data, including data that were not available when the operator made the daily decision (such as influent BOD, effluent BOD, or F/M ratio). Based on this information, the operator can objectively judge the quality of his decisions against the expert system's and, over time, continually improve and refine the knowledge base.

As additional and more complex rules are collected and stored in the system, it is possible that the expert system may suggest actions that are contradictory. For example, one triggered control rule might suggest an increase in variable A while another rule suggests a decrease. To help the operator resolve conflicts, the tool box can generate a statistical report that gives the number of times in the past that an operating condition occurred, the dates when it occurred, relevant operational data, and the actions that actually were taken. The operator can use this report to resolve the conflict. Study of such reports may also lead to modifying the existing rules or adding new hierarchical rules to the expert system's knowledge base.

Table 1. A Portion of a Performance Report

Date	BOD mg/L	Temp °F	MLSS mg/L	D.O. mg/L	RAS Ratio	WAS kGal	Suggested Action	Actual Action
06/07/85	13	65	1294	2.4	0.58	5.0	none	none
06/08/85	18	65	1266	2.5	0.57	5.0	none	none
06/09/85	13	66	1510	4.2	0.57	14.5	1	1
06/10/85	12	66	1354	3.2	0.57	5.0	8	8
06/11/85	12	66	1258	2.7	0.58	5.0	none	none
06/12/85	23	66	1294	3.4	0.57	5.0	none	none
06/13/85	28	65	1376	4.2	0.57	5.0	none	none
06/14/85	30	68	1424	1.5	0.57	15.4	1	1
06/15/85	12	68	1302	4.2	0.57	5.0	8	8
06/16/85	12	67	1330	2.6	0.57	15.4	1	1

The use of the tool box is based on the idea that an expert system should always be considered capable of being improved. The strategy for progressive refinement involves an interactive process between the expert system module and competent plant operators. The development process might proceed as illustrated in Figure 2. The curve depicts the true (but unknown) relationship between two monitoring variables, $MV(1)$ and $MV(2)$, and a control action X. At an initial stage, experts might do no better than the approximate rule, shown in the top panel of Figure 2:

IF $MV(1) > 10$ and $MV(2) > 2500$ THEN take action X.

Continued evaluation of the rules should reveal how step-wise approximations can be added so the ES will more closely approach the true response curve.

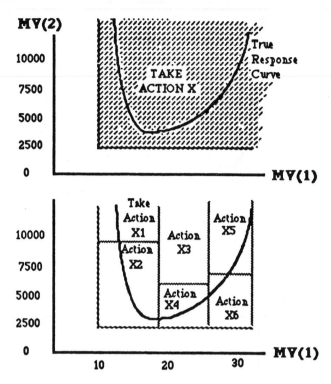

Figure 2. Progressive Improvement of the Rule Base

4. An Approach to Knowledge Acquisition

The following steps were taken to develop and check the rule base. A general examination of the plant performance was made using historical data and interviews with the supervising, operating, and laboratory staff. An historical data base was constructed and analyzed to select representative time periods and to construct test problem sets. Simulation experiments were used to derive the appropriate definitions for the fuzzy terms involved in the rules. These definitions become the control action levels in the rule base. The refined rules were then tested using sets of test problems which are independently evaluated by the expert operators.

The plant engineer and his most experienced and skilled staff were considered to be the domain experts in this study. They were identified by examining their past performance at the plant, their operator classification grade (which is based on a series of tests, training courses completed, and experience), and by their reputation within the water pollution control community.

Control rules were elicited by having the expert operators evaluate case study problems. The procedure was to (a) prepare a number of problem cases from the plant's daily data records, (b) present these problem cases one at a time to the expert operator, (c) collect, summarize, and codify the heuristic rules used by the operator to handle each case, and (d) store these rules in the knowledge base so they could be quantified, tested, and refined.

Control rules collected from the domain experts usually contained fuzzy terms such as high and low, and these had to be given quantitative site-specific definitions. Then the rules needed to be verified, refined, and probably modified before an expert system was implemented for regular use. Reference distributions (Berthouex and Hunter 1983) for the variables involved in the control rules were used to select trial settings for the action levels (i. e., the high and low levels) of the fuzzy variables. These trial settings were tested by simulation in order to determine the action limit settings that gave the fewest incorrect process control decisions, as defined by the performance index. The derived settings were used in the expert system to evaluate its performance in diagnosing process conditions and suggesting control actions on a set of problem cases. The selected problem cases were further analyzed to discover discrepancies between the operator's decisions and the advice offered by the expert system. This analysis indicated that some rules needed modification, that new rules had to be added, or that conflicts between rules needed to be resolved.

5. Evaluating the System's Accuracy

The objective was to examine the system's accuracy in diagnosing operating conditions and in providing appropriate control actions for solving problems. Accuracy was measured by comparing the expert system's suggestion with expert operator's advice. In some cases it is be difficult to objectively define a single correct answer for a given problem case. Therefore, a suggestion generated by the expert system was considered "correct" if it was acceptable to the domain expert. Accepting a domain expert's decision as a datum for comparison is a subjective, but commonly used, method for assessing the performance of an expert system especially when the expert system is in the prototype stage (Gaschnig at al. 1983, Yu 1979).

A set of problem cases selected from the treatment plant's daily data records was prepared for the evaluation. For each problem case, the "expert" operator was asked to identify the process control problem and to recommended control action for handling the problem. The same problem cases were analyzed by the ES and the suggestions generated were compared with the operator's answer. When disagreements were noted, the operator was asked to assess the correctness (acceptability) of the ES recommendation. The degree of accuracy was computed as the percentage of cases where the ES suggestion agreed with or was accepted by the expert operator.

6. A Case Study

The Fort Atkinson Wastewater Treatment Plant is an activated sludge treatment plant with an average design capacity of 2.7 mgd (10.2 m³/d). Normally the plant is operating at approximately three fourths of the average design flow, although each year during certain periods, especially spring snow melt, the daily influent flow may become as high as 8 mgd (30.3 m³/d). The raw sewage flow exhibits a weekly pattern, with weekday loadings being greater than weekends. This is due in part to industrial loading on the plant. The plant almost always meets its final effluent standards of 30 mg/L for both BOD and SS. The principal variables that were routinely recorded at this plant were mixed liquor suspended solids (MLSS), waste sludge solids concentration (WSSC), hydraulic detention time (HDT), sludge blanket depth (SD), and sludge age (SA). The principal control variables were waste activated sludge flow rate (WAS) and return activated sludge flow rate (RAS).

Rule Elicitation. Two operators at this plant were considered experts. They each analyzed one hundred problem cases constructed from past daily operating records. For each problem case, the data for several previous days were provided so they could review the recent process history. The rules elicited from the experts for control of the activated sludge process were:

1   IF mixed liquor suspended solids (MLSS) is high and increasing
     THEN increase waste activated sludge flow rate (WAS).
2   IF MLSS is low and decreasing
     THEN decrease WAS.
3   IF raw influent sewage flow is high and hydraulic detention time (HDT) is low
     THEN decrease return activated sludge flow rate (RAS).
4   IF raw influent sewage flow is low and HDT is high
     THEN increase RAS.
5   IF MLSS is normal but decreasing and waste sludge solids concentration (WSSC) is low and decreasing
     THEN decrease WAS.

6   IF MLSS is normal and WSSC is normal and Sludge Age (SA) is high
        THEN increase WAS.

7   IF MLSS is normal and WSSC is normal and Sludge Age (SA) is low
        THEN decrease WAS.

8   IF raw influent sewage flow is normal and HDT is high
        THEN increase RAS.

9   IF raw sewage flow is normal and HDT is low
        THEN decrease RAS.

10   IF Sludge Depth (SD) is high and increasing
        THEN increase RAS.

11   IF MLSS is normal but increasing and WSSC is high and increasing
        THEN increase WAS.

One aspect of the rules that needs explanation is the use of both flow rate and hydraulic detention time (HDT) in some of the rules. This plant has multiple parallel basins and sometimes basins are taken out of service for various reasons. It is possible, therefore, to have low HDT without having extraordinary high flow.

Before going on to assess the performance of this rule base it should be stated that eleven basic rules do not fully describe an activated sludge process. It is not expected to deal with all possible operating conditions. For example, it does not make explicit decisions about suppressing sludge bulking, nitrification, or recovery from a toxic shock. It did, according to the expert operators, seem sufficient to describe the operating situations represented by the test data set.

The ES also contains rules for diagnosing and moderating sludge bulking, but these were taken from Jenkins et al. (1986). Since they were not elicited from the operators, nor tested during this project, they are not discussed. Nitrification and toxic shock conditions did not exist at this plant so no rules were developed for these situations, but such conditions can be included in an ES.

Rule Calibration. These control rules were tested by simulation to learn whether they could be improved. Simulation, here, means running the proposed ES rules on historical data that were selected to represent conditions of special interest. This started with construction of reference distributions constructed using data from periods when the process was under control. The upper and lower 5% levels, and a few other percentile levels, for these distributions are given in Table 2. These percentile levels are used to propose trial settings for the 'high' or 'low' conditions for the variables that are to be tested in simulation runs.

Table 2. Selected Percentiles for the Key Variables

Variable	95%	85%	75%	50%	25%	15%	5%
MLSS	2220	1892	1768	1500	1353	1284	1130
WSSC	9617	8083	6900	5967	5033	4567	3817
SA	8	6	5	4	2	2	1
SD	3.25	2.75	2.25	2.00	2.00	2.00	1.00
HDT	5.1	4.5	4.1	3.9	3.4	3.1	2.3
RSF	3.4	2.4	2.1	1.9	1.75	1.7	1.4

The simulation experiments investigated each decision variable one by one. Each simulation run evaluated 100 test problems. For example, the first section of Table 3 shows the MLSS settings finally selected after evaluating the 100 test problems with a variety of high-low settings of the action limits. Settings for MLSS were evaluated while the levels of all other variables were held constant. The best settings for MLSS were maintained for the next simulation trial which investigated settings for waste sludge solid concentration (WSSC). In these trials, the high-low levels for WSSC were varied while the settings for all other variables were fixed. A similar one-variable-at-a-time investigation was made for each variable.

Each simulation trial was characterized by the number of disagreements and the performance index. The performance index was calculated with the utility function given earlier assuming that misses and false alarms were equally important, so $W1 = W2 = 0.5$. A "miss" was counted when the human expert recommended taking action but the expert system did not. Since the ES sometimes recommended more than one action, there could be more than one miss on a given day. A "false alarm" was counted when the computer recommend taking action and the human expert thought action was not required. The best settings for the control variable were taken as those giving the lowest performance index value. Table 3 shows how the performance index improved as the setting of the variables were adjusted.

Table 3. Results of Simulation Trials to Define the Low/High Action Levels for Fort Atkinson.

	Trial 1	Trial 2	Trial 3	Trial 4	Trial 5	Trial 6
MLSS Action Levels	1219/1814	1219/1814	1219/1814	1219/1814	1219/1814	1219/1814
WSSC Action Levels	3817/8600	3817/6900	3817/6900	3817/6900	3817/6900	3817/690
SA Action Levels	2/8	2/8	2/8	2/8	2/8	2/8
SD Action Levels	1.50/3.25	1.50/3.25	1.50/3.25	1.50/2.50	1.50/3.25	1.50/2.50
HDT Action Levels	2.2/5.1	2.0/5.1	2.0/5.1	2.2/5.1	2.9/5.1	2.9/5.1
RSF Action Levels	1.4/3.4	1.5/3.4	1.4/3.4	1.4/3.4	1.4/3.4	1.4/2.4
Misses	22	22	22	5	2	2
False Alarms	5	4	4	10	12	12
Perf. Index	13.5	13.0	13.0	7.5	7.0	7.0

The calibration levels presented in Table 3 were determined by investigating the decision variables one-by-one in a certain sequence (MLSS, WSSC, etc.). To learn whether this affected the outcome, the exercise was repeated using several different sequences of variable manipulation. There was no significant difference in actions levels reached with the different sequences. This ad hoc one-by-one process of testing the decision variables seemed to work, but the implication that one variable can be studied independently of the others is strongly counter-intuitive to the way most engineers understand the treatment process. Drawing an analogy to physical experimentation would also suggest that this approach is risky, since manipulating factors one by one in a physical experiment can be inefficient and misleading. It is well known to be less sound than a two-level factorial experiment (Box et al. 1978). Future experience may indicate that a numerical factorial experiment is indeed more likely to find the best settings, and to do so with fewer simulations. This remains an area for further investigation.

Verification of the Calibration. Based on the derived definitions, the expert system's performance on the 100 problem cases were analyzed. Table 4 summarizes the 12 cases of disagreement between the expert system and the operator that were discovered.

Table 4. Analysis of Unmatched Decisions in the Simulation for Fort Atkinson.

Case	Problem Identified by ES[1]	Action Recommended by ES[2]	Action Taken by Operator	Misses	False Alarms
1	3,10	4,3	3	0	1
2	2,3,10	2,4,3	2,3	0	1
3	none	none	2	1	0
4	6,10	1,3	3	0	1
5	10	3	none	0	1
6	1,3	1,4	none	0	2
7	3,10	4,3	none	0	2
8	6,10	1,3	1	0	1
9	3	2	none	0	1
10	10	3	none	0	1
11	2	2	2,3	1	0
12	10	3	none	0	1

[1] The problem identification numbers correspond to the rule numbers given in the text.
[2] Action codes: 1 = increase WAS, 2 = decrease WAS, 3 = increase RAS, 4 = decrease RAS.

A case was considered in disagreement if diagnosis of the operating condition and the suggested control action generated by the expert system was not identical to that of the expert operator. The "problem identified by the ES" is the rule which the ES invoked to generate a control recommendation. These rules were listed earlier. In several instances, more than one rule was activated. The expert system missed suggesting control actions twice, and gave false alarms 12 times.

Analysis of Disagreements. Much can be learned by discussing the cause of disagreements. Several general guidelines for resolving conflicts were established in this way. These are mentioned below.

• When conflict arises, select the control action that has the most immediate effect on protecting effluent quality. In disagreement 1 (Table 4), the problems in the treatment process identified by the ES included high influent flow and high sludge blanket depth in the final clarifier. Two expert operators agreed with the condition diagnosis. The ES triggered two contradictory rules:

Rule 3   IF raw influent sewage flow (RSF) is high and HDT is low
            THEN decrease RAS.
Rule 10   IF Sludge Depth (SD) is high and increasing
            THEN increase RAS.

The experts elected to increase RAS in order to reduce the high sludge blanket depth in the final clarifier. They noted that sludge depth had been high for several days and that not taking action now was likely to cause an immediate deterioration in effluent quality because of solids carryover. They preferred the action having the most immediate and direct link with effluent quality.

• When conflict arises, examine trends in the variable to determine the appropriate action. In disagreement 2, the problems identified by the expert system included low and decreasing MLSS, high raw influent flow, and high sludge depth. This triggered three control rules and a conflict exists between rules 3 and 10:

Rule 2   IF MLSS is low and decreasing
            THEN decrease WAS.
Rule 3   IF raw sewage flow (RSF) is high and HDT is low
            THEN decrease RAS flow rate.
Rule 10   IF Sludge Depth (SD) is high and increasing
            THEN increase RAS.

The operators had recommended a decrease in WAS because of the low and decreasing MLSS, and an increase in RAS because of the high sludge depth on the clarifier tanks. These two suggestions matched the advice generated by rules 2 and 10. The problem of high influent flow was acknowledged by the experts, but the flow had been decreasing for the past several days so they felt that the problem was being naturally eliminated and that action was not needed to counter this factor.

• Recent control actions may influence today's decision. In disagreement 3, the experts felt that the MLSS was low and decreasing and they decided to decrease the WAS. The expert system considered the MLSS level to be within the normal range and therefore suggested no control action. The ES decision of "no action" was acceptable to the operators because taking action one day later would cause no harm. The operators pointed out that they had increased the WAS four days earlier when the MLSS was 1966 mg/L, after which the value of MLSS had fallen to around 1300 mg/L and was still decreasing. They now wanted to take action to prevent the condition of the MLSS from becoming even lower.

• Beware of acting on the basis of a single unusually high or low value. Several case problems illustrate this heuristic. In disagreement 4, the ES recognized high sludge age and high sludge depth, which triggered two rules:

Rule 6   IF MLSS is normal and WSSC is normal and SA is high
            THEN increase WAS.
Rule 10   IF Sludge Depth (SD) is high and increasing
            THEN increase RAS.

The operators wanted to increase the RAS only, feeling that the high sludge age for this for one day was not truly indicative of process conditions. They wanted to observe it further before making a decision to lower the sludge age. The true sludge age of a process should not change rapidly, but the computed values can be quite variable. A single high value can easily result from measurement error in one of the measured values used in the computation of sludge age. This is a situation where a moving average plot is helpful in smoothing the data and reducing the chance for over-reacting.

In disagreement 5 the triggered rules, which give conflicting advice with respect to RAS adjustment, were:

Rule 1   IF MLSS is high and increasing THEN increase WAS.
Rule 3   IF raw sewage flow is high and HDT is low
              THEN decrease RAS flow rate.
Rule 10   IF Sludge Depth (SD) is high and increasing
              THEN increase RAS flow rate.

The suggestion of increasing the WAS was acceptable to the operators. The operators took no action to change RAS, noting that sludge depth and raw influent flow had both increased suddenly. Lacking evidence of a trend, they felt no risk was attendant with waiting to see how conditions would develop. Moving averages plots are also useful in cases like this one.

In disagreement 8, 10, 11, and 12, the ES identified high sludge age and high sludge depth as problems and this triggered two rules:

Rule 6   IF MLSS is normal and WSSC is normal and SA is high
              THEN increase WAS flow rate.
Rule 10   IF SD is high and increasing
              THEN increase RAS flow rate.

The operators elected to increase the WAS since the sludge age had been high for several days. However, since the value of the sludge depth had increased suddenly they preferred to wait before deciding to increase RAS in order to lower the sludge depth.

• When conflict arises, sometimes there is little or no risk to wait until more data are collected. In disagreement 5 the ES advice to increase the RAS was acceptable to the good operators, even though it did not match their decision, but they decided to take no action because they had made this adjustment on the previous day. The sludge depth remained high but was not increasing so the operators wanted to postpone action in order to avoid a possible over adjustment.

In case 7 the ES identified two problems identified and contradictory advice resulted The diagnosis of high raw sewage flow and low HDT generated a suggestion to decrease the RAS. The simultaneous diagnosis of high sludge depth activated a suggestion to increase the RAS. The operators did not disagree with the diagnosis of the expert system, but they decided to take no immediate action, deciding that both the raw influent flow and the sludge depth seemed to show an increasing trend and it wasn't clear that either increasing or decreasing RAS would be beneficial. In fact, they feared that either action might make conditions worse, but that taking no action carried no immediate risk. Therefore, the operators decided to wait and see how conditions would develop.

Rule Modifications. The analysis contributed several findings about the heuristic rules of the good operators. This led to modification of some original rules as well as the addition of some new rules to the expert system. It may not be the most optimal long-term strategy to modify generally successful rules in an attempt to eliminate a small number of discrepancies. We may be ahead merely to have the expert system bring contradictory suggestions to the attention of the operator, and then to ask the operator to select a policy. This is completely within the spirit of the ES approach since the goal is not to eliminate the operator, but rather to free him of the routine decisions so that he may concentrate his creativity and intelligence on the special control problems. Nevertheless, to illustrate how modification might be made, and also to investigate whether gains were at all possible, some modifications and additions were made and tested.

The major change was to incorporate moving averages in some control rules, especially the rules associated with raw sewage flow (RSF), sludge age (SA), hydraulic detention time (HDT), and sludge depth (SD). This reduces the influence of a single unusual value and consequently reduces the tendency to over-adjust the treatment process. Five-day moving averages were used arbitrarily for all variables. This was done without trying to select the best kind of moving average for each variable even though one might expect that, for example, a seven or ten-day moving average might be best for sludge age while a three-day average might be suitable for another variable. If an unsuitable average is used, a lag effect is created and this could cause a degradation in the ES performance. Additional work is needed to find the most appropriate kind of moving average for each variable.

Expert System Accuracy and Effectiveness. The ES, with the modified rules and their selected action levels, was evaluated with respect to accuracy in diagnosing operating problems and in suggesting appropriate control actions for solving these problems. The diagnoses and suggestions of the ES were compared with those obtained from two expert operators for a new set of 120 case problems. The results of the evaluation regarding the ES's accuracy showed that the problem identified by the expert system was acceptable to the expert operators in 118 of the 120 cases (98.3%). The control action suggested by the system was acceptable to the operators in 117 of the 120 cases (97.5%). This level of success cannot be predicted for all conditions, since it is likely that the full range of operating problems has not been encountered in the test set of 120 problems cases, but the results do indicate that a working rule base can be developed using the approach outlined here.

Improving Operator Decision Making. One reason for building an expert system is to allow human expertise to be shared. The ES was evaluated with respect to improving the decision making performance of the less experienced, and presumable less skilled, operators. The set of 120 problem cases that had been evaluated by the expert operators was divided into two similar sets of 60 problems. Three less experienced operators evaluated one set of 60 problems without assistance, and the other set with the assistance of the ES. "Assistance" means that the less skilled operators were given the problem case plus the ES diagnosis and suggestions. Table 5 presents the results of the evaluation regarding the system's effect in improving the less skilled operator's performance.

Table 5.ˑ Effectiveness of the ES in Improving Operator's Performance.

Operator	Status	Acceptable	Unacceptable	% Correct	Penalty
A	without ES	47	13	78	54
	with ES	54	6	90	22
B	without ES	48	12	80	49
	with ES	53	7	88	25
C	without ES	47	13	78	54
	with ES	56	4	93	13

The operator's performance for each set of cases was measured as the percentage of his answers that matched or was acceptable to the expert operator. A second measure of performance was a penalty score, based on a scale ranging from ten for a serious error to zero for a tolerable error. Without the assistance of the system, in 13 of the 60 cases (22%) the answers given by operator A were not acceptable to the good operators and resulted in a penalty score of 54. The penalty score reflects the number of misses and also the severity of making the wrong decisions. Its computation was described earlier. With the assistance of the ES, operator A gave only 6 out of 50 (10%) answers that were not acceptable to the good operators and resulted in a penalty score of 22. With assistance from the ES, operator A was able to reduce his "error" by half, 6 unacceptable decisions compared to 13. In terms of the penalty score, this is an improvement of 32 points. Similar degrees of improvement were shown by operators B and C. On average, the less skilled operators reduced their wrong decisions by half with the assistance of the system.

7. Summary and Conclusions

The trial applications of the STATEX management information system indicate that it can provide benefits to the operator. The expert system, even though offering many areas for future development, does seem to have potential for improving the decisions of less skilled operators.

A case study method was used for interrogating "expert" operators at treatment plants to elicit control rules. This approach for rule elicitation was found to be cognitively simple and natural for the operator. This is not surprising since it uses data records from his plant as the problem cases for the study. The first step is to collect control rules, which are structured in qualitative or fuzzy terms. The advantages of keeping fuzzy terms in rule syntax are (a) ease of elicitation – it is generally easier to elicit rules from if operator is required merely to categorize, and (b) extensibility – such rules can be programmed into the expert system and be retained for use at all or most plants, whereas this would not be the case if numerical values were to be specified.

The simulation utility provided in the tool box was useful to refine and extend the general rules in light of the operators' own experience and local conditions. Empirical reference distributions and simulation are used to aid in deriving the most appropriate numerical definitions for the fuzzy terms in the control rules. The performance report provided by the tool box has also shown to be helpful in extending or modifying the rules collected in the first stage and in resolving conflict among the rules. Deficiencies in these rules will be identified with the help of the tool box utilities and the knowledge base will evolve and improve as the system is used.

The syntax of fuzzy rules may offer some general applicability, but there will be varying interpretation of the qualitative terms in the rules from plant to plant. A means of replacing these fuzzy terms with appropriate site-specific quantitative levels has been developed and demonstrated in the application. It involves construction of reference distributions, simulation runs of selected problem cases, and utilization of performance reports provided by the tool box. Although the results of the simulation did not seem significantly related to the sequence of the variables, they are critically dependent on the quality and number of selected problem cases. The set of problem cases used for the simulation trials should contain as many operating conditions as possible. Failure to find typical problem cases may result in failure to derive the most appropriate definition for the variable, or it could mean that a necessary rule is never discovered during the development process. These problems are not serious since records kept by the tool box during regular use of the ES can be checked from time to time to assess weaknesses in the current knowledge base.

An important principle in controlling the treatment process is to avoid over-adjustment, since this reduces rather than improves stability. Good operators seem to intuitively understand this concept. This is reflected in several of their strategies: (1) They do not rely only on current values but also data from the previous few days so as to obtain a view of trends and a whole picture of the process. Accordingly, they might decide to take more or less action in order to avoid over-adjustment of the process. (2) Operators usually make the adjustment in a step-by-step manner rather than making a large step change. This preference allows them to be comfortable with the way the ES works, since it only suggests "increase" or "decrease" and does not advise on how strong a change should be made. (3) The effect of a control action may last for several days or take several days to become effective. Hence, knowing that similar action was recently taken may affect operators' decision for the current day. (4) Operators tend not to jump to conclusions on the basis of a single suddenly abnormal value of the variable. Most of the time they regard such values as sampling or measuring error. When this is suspected they prefer to collect more data before making a decision.

The ES can generate more than one recommendation for making process adjustments, and these can be contradictory. The operator must resolve the conflict, since the ES contains no hierarchical rules for doing so. If taking no action seems a safe option, the operators usually choose to wait until more information becomes available. If delay seems risky, they tend to select the action that has the most immediate effect on improving the quality of the effluent, or the action that does the least potential harm. Examining trends of key variables is often helpful in determining which control action is needed.

8. Acknowledgements

The tests of the expert system required many hours of interviews with plant operators to encode their operating rules and work through the case problems. We are grateful for the generous cooperation of Roger Sherman, Paul Christensen, Terry Vaughn, Bill Fredrickson, and Venard Cypher.

Although the research described in this article has been funded in part by the United States Environmental Protection Agency under assistance agreement CR-812655-01-0 to The University of Wisconsin-Madison, it has not been subjected to the Agency's peer and administrative review and therefore may not necessarily reflect the views of the agency, and no official endorsement should be inferred.

Appendix I. References

Beck, M.B., Latten, A., and Tong, R.M. (1978). "Modelling and operational control of the activated sludge process in wastewater treatment." Report: International Institute for Applied Systems Analysis, Laxenburg, Austria.

Berthouex, P.M., and Hunter, W.G. (1983). "How to construct reference distributions to evaluate treatment plant effluent quality." *J. Water Pollut. Control Fed., 55*, 1418-1424.

Berthouex, P.M., Lai, W., and Darjatmoko, A. (1989). "Statistics-Based Approach to Wastewater Treatment Plant Operations." *J. Envir. Engr. Div., ASCE, 115*, 650-671.

Box, G.E.P., Hunter, W.G., and Hunter, J.S. (1978). *Statistics for Experimenters: an Introduction to Design, Data Analysis, and Model Building.* Wiley, New York, NY.

Gaschnig, J., Klahr, P., Pople, H., Shortliffe, E., and Terry, A. (1983). "Evaluation of expert systems: issues and case studies." *Building Expert Systems*, Hayes-Roth F., et al., Ed., Addison-Wesley, Reading, MA, 241-280.

Huang, M.S., Shenoi, S., Mathews, P.A., Lai, S.F., and Fan, T,L. (1986). "Fault diagnosis of hazardous waste incineration facilities using a fuzzy expert system." *Expert Systems in Civil Engineering*, C.N. Kostem and M.L. Maher, Ed., ASCE, New York, NY, 30-37.

Jenkins, D., Richard, M.G., and Daigger, G.T. (1986). *Manual on the Causes and Control of Activated Sludge Bulking and Foaming.* Ridgeline Press, Lafayette, CA.

Johnston, D.M. (1985). "Diagnosis of wastewater treatment processes." *Proceedings, Computer Applications in Water Resources*, ASCE, New York, NY, 601- 606.

Kramer, M.A. (1987). "Malfunction diagnosis using quantitative models with non-boolean reasoning in expert systems." *AIChE Journal, 33*, 130-140.

Lai, W. (1989). "A Statistics-based Information and Expert System Approach for Wastewater Plant Control." Ph.D. Thesis, University of Wisconsin-Madison.

Maeda, K. (1984). "A knowledge based system for wastewater treatment Process." *Proc. of 9th IFAC*, 6, 3251-3256.

Maeda, K. (1985). "An intelligent decision support system for activated sludge wastewater treatment process." *Instrumentation and Control of Water and Wastewater Treatment and Transport Systems*, Drake, R.A.R., Ed., Advances in Water Pollution Control, Pergamon Press, London, 629-632.

Tong, M.R., Beck, B.M., and Latten, A. (1980). "Fuzzy control of the activated sludge wastewater treatment process." *Automatica, 16*, 695-701.

Vassos, T.D. (1987). "Development of an Activated Sludge Process Control Strategy Using Bayesian and Markovian Decision Theory." Ph.D. Thesis,

University of British Columbia.

Vitasovic, Z., and Andrews, J.F. (1987). "A rule-based control system for the activated sludge process." *Systems Analysis in Water Quality Management*, IAWPRC, Pergamon Press, New York, NY, 423-432.

Yu, V.L. (1979). "Evaluating the performance of a computer-based consultant." *Computer Programs in Biomedicine, 9,* 95-102.

Appendix II. Glossary

Activated Sludge (AS) Process - a wastewater treatment process that mixes raw wastewater, flocs of bacterial cells, and air together in a reactor vessel called an aeration basin. The bacteria consume organic compounds in the wastewater and are then separated from the treated water by gravity settling in a second basin called the final clarifier. A portion of the settled bacterial solids is recycled and mixed in the aeration basin with the incoming raw wastewater.

Biochemical Oxygen Demand (BOD) - the potential of a wastewater to consume oxygen through biological reactions; a gross measure of the organic strength of a wastewater.

Hydraulic Detention Time (HDT) - the average time that wastewater is oxygenated and exposed to bacterial treatment in an activated sludge aeration basin.

Mixed Liquor - the contents of the aeration basin; a mixture of raw wastewater and concentrated bacterial sludge that is recycled from the final clarifier.

Mixed Liquor Suspended Solids (MLSS) - the concentration (mg/L) of bacterial floc solids present in an activated sludge aeration basin.

Return Activated Sludge (RAS) - the flow of bacteria-rich sludge that is recycled from the bottom of the final clarifier to be mixed with raw wastewater in the aeration basin of an activated sludge process.

Sludge Age (SA) - the average length of time that bacterial cells remain within the activated sludge process before being lost or deliberately wasted form the system.

Sludge Blanket Depth (SBD) - the depth of the layer of settled and concentrated bacterial solids that resides in the bottom of the final clarifier, and from which return sludge is pumped.

Suspended Solids (SS) - the dry weight of solid particles in a unit volume of wastewater that can be captured by filtration on ordinary filter paper.

Waste Activated Sludge (WAS) - the portion of settled bacterial sludge that is removed from an activated sludge process in order to control the amount of bacteria held within the treatment system.

CHAPTER   8

Knowledge Acquisition for Postearthquake Usability Decisions

Zahra-El-Hayat Tazir
Tommaso Pagnoni
Carlo Gavarini

1. Introduction

The case study presented in this chapter is in the area of earthquake engineering. It addresses the emergency damage and usability assessment of buildings after an earthquake. The damage assessment consists in the quick survey of the buildings in the area hit by an earthquake with the principal objective of assessing their safety. The usability evaluation involves recommending whether the surveyed buildings are fit for continued use and occupancy. Usually, this operation results in posting the building as either habitable, unsafe, or to be reinspected.

The main purpose of this operation is to save human lives and prevent injuries. These functions are accomplished by identifying and limiting access to buildings that have been seriously weakened by the earthquake and are therefore threatened by aftershocks. Another objective is to reduce the economic cost and human suffering caused by the disaster by identifying habitable and easily reparable buildings (Anagnostopoulos et al., 1985; Gavarini, 1985a, 1985b).

After a moderate or strong earthquake strikes a populated region, a large number of buildings suffer various degrees of damage; they need to be inspected and their safety evaluated. However, because the demand for qualified building inspectors is far greater than their number, inexperienced engineers and insufficiently trained technicians are assigned to this difficult task with no specific criteria as to what to do or how to make usability decisions (Applied Technology Council, 1989a; Gavarini 1985a, 1985b). The result is that a large number of these decisions are generally conservative. This translates into many buildings being classified as either needing an additional inspection or as unsafe; consequently, a return to normal life is considerably delayed. Also, inspectors who lack experience and knowledge may reach decisions that endanger people and property. Finally, the lack of specific guidelines may result in nonuniformity of the usability decision; eg., different inspectors may reach conflicting decisions regarding the safety of the the same building (Anagnostopoulos, 1985; Applied Technology Council, 1989a; Gavarini 1985a, 1985b).

The problem is the lack of availability of the specialized knowledge required to make usability decisions on buildings after an earthquake. Its solution involves a twofold procedure: first, the specialized knowledge must be collected and organized; then, and just as important, this knowledge should be presented to the novice inspector in a flexible, interactive, and understandable format. The first task consists in acquiring the knowledge from domain experts and structuring it. The second point can best be addressed through knowledge-based technology. Indeed, the transfer of knowledge through mere questionnaires and manuals has proven to be relatively inefficient in this area, mainly because of time constraints due to emergency conditions. Another important reason for this inefficiency is that the nature of the expertise involved is to a large extent heuristic (Applied Technology Council, 1989a).

To remedy this situation, the authors propose the use of a Knowledge-Based System (KBS) to transfer the knowledge of experienced engineers (domain experts) to the less experienced operator, and make it available on the site of the inspection. To this end, Amadeus[*], a KBS whose purpose is to help make on-site usability decisions on buildings after an earthquake, was developed (Pagnoni et al., 1989). It is an advisory system which (a) guides the engineer through the evaluation by suggesting steps to take during the inspection and, (b) proposes, based on the data gathered by the operator, a decision regarding the usability of the building. The methodology implemented in this KBS has been developed in Italy and relies, for the most part, on data and observations resulting from earthquakes in that country. However, because similar problems occur in any other seismically active area, the methodology should still be applicable, with some minor modifications, in any other such region.

The project started in the spring of 1987. After about three months, a simple demonstration prototype was developed. The restructuring, validation through interaction with the expert, refinement and addition of domain knowledge, as well as the development of a database interface, took another two years.

This chapter will examine the knowledge acquisition (KA) process involved in the development of Amadeus. In general, such a process involves two principal stages. The first combines eliciting, analyzing, and interpreting the knowledge that experts use when solving a given problem. In the present case this task has been resolved through literature reviews and interviews with the domain expert who authored the current postearthquake evaluation guidelines in use in Italy. The

---

[*]Advisory Methodology for the Assessment of Damages after Earthquakes and Usability of Structures.

second step involves transforming the knowledge acquired into a suitable machine representation. Amadeus has been developed in PersonalConsultant[Plus], a Lisp-based expert system development tool, which incorporates a basic knowledge acquisition module (PersonalConsultant[Plus]). The use of an existing shell has greatly facilitated this second stage.

In what follows, we will first review the problem and the suitability of expert system technology for this case. Next, we will address more specifically the knowledge acquisition process and describe how it was approached. In that section, the various other aspects of knowledge engineering, in particular its knowledge representation (KR) facet, and the influence it had on the KA process of the project, will be examined. Finally, some general evaluation of the work will be presented together with an appraisal of the methods used.

2. Background

Problem Description

Good management of the emergency postearthquake damage and usability evaluation process is crucial to the recovery of the distressed areas. Unfortunately, it is difficult to train a sufficient number of building inspectors immediately after an earthquake strikes (Applied Technology Council, 1989a ; Gavarini, 1985a). Moreover, the assessment requires much expertise and heuristics. Indeed, without a fair amount of experience it is difficult to accurately estimate the state of damage of most buildings. In addition, most of the existing damage assessment procedures rely heavily on the use of subjective scales and on the judgment of the inspector (Applied Technology Council, 1989a). The fears associated with being nonconservative in this judgment lead most of the inexperienced operators to be overly cautious. Consequently, they classify many buildings as either unsafe or requiring additional inspections. Indeed, it is not rare that these inspectors reach the conclusion *"unable to classify, reinspection recommended."* (Anagnostopoulos, 1985). The result is that a number of habitable buildings are incorrectly classified. These buildings have to remain closed to occupancy until more detailed inspections take place. This situation creates much strain on an already seriously distressed environment. Useful resources are wasted and more shelters than needed are used.

The problem is therefore severe and needs to be seriously addressed. In the last fifteen years, because of the potentially dramatic consequences that can result from mismanagement, local authorities and government agencies in areas prone to earthquakes have paid closer attention to this operation. A review of the available methods shows the following: The forms the inspectors use to record their observations, recommendations, and usability decisions, rely heavily on engineering knowledge and judgment. Indeed, most forms require subjective judgment on the part of the inspector. Moreover, the format of these questionnaires is more tailored toward ease of data input for later processing purposes than toward immediate ease of use by the damage inspector (Applied Technology Council, 1989a ; Pagnoni *et al.*, 1989). More recently, in the US, the Applied Technology Council took on a project whose aim was to establish guidelines for postearthquake damage assessment. Aware of the shortcomings of the existing methods, the new methodology attempts to address them. The project resulted in two documents: a report describing the methodology, the ATC-20, and a field manual for on-site use by the inspector, the ATC-20-1 (Applied Technology Council, 1989a and 1989b). The improvements provided through this methodology are notable.

Because of the importance of the emergency postearthquake usability decision, an earlier attempt was made to formulate a methodology that could capture and make perceptible to a potential user a coherent reasoning process leading to the usability evaluation. This endeavor aimed at rationalizing and structuring the mental process through which assessments could be reached. The formulated methodology could then guide potentially inexperienced inspectors in the assessment (Gavarini, 1985a, 1985b).

In that study, a series of guidelines in the form of rules, a flowchart, and decision tables were developed to help the inspector in making the usability assessment (Gavarini, 1985a). The author of these guidelines was the expert for the project discussed herein, and the resulting methodology was the basis for the development of Amadeus. The guidelines proposed by the expert were the result of his involvement with postearthquake situations for many years. They derive from the experience he accumulated through these events, and from his observations and interactions with colleagues.

The purpose of the guidelines is to articulate and clarify the decision process to be followed. They were designed so as not to restrain the user's freedom or ability to think and decide independently. Above all, they emphasize the important points of the process. This attempt was the first step in the KA process for this project.

Objective of the KBS

Again, the principal objective of Amadeus is to guide the inspectors in their reasoning by identifying the various steps of the evaluation. It also directs their attention to the relevant facts, and suggests, based on the gathered information, a final decision with respect to the habitability status of the building under inspection. It should be emphasized that Amadeus is by no means a replacement for the inspectors, but is merely their knowledgeable assistant. According to the expert's model of the domain area, the KBS points out the elements of interest to the decision making process in each special case, ignores the irrelevant factors, and suggests a qualitative result for each of the required tasks. Eventually, these results combine to propose a usability decision for the buildings.

However, the final decision should still be the inspectors'. Therefore, to give them more freedom of decision and acknowledge their possible competence, the system has been designed so that each major conclusion resulting from its knowledge-base can be altered by the inspector. At the end of the evaluation of each primary task, the user is asked if the result is satisfactory. He or she can query the system as to why such an outcome resulted from the observations. If the inspectors are confident about their judgment and can take into account elements of reasoning that were unavailable to the system, they can modify the result of the evaluation, and the system then continues its reasoning with the new value for the particular task. The inspector may also disagree with the final conclusion of the evaluation suggested by the system.

Another objective of the system is to help in the management of the emergency situation by gathering all the relevant data. The resulting database could also be useful for future risk studies. Finally, the system could also be used as a means to train volunteers so they become familiar with the domain, think correctly about the problem, and learn the important aspects of the operation.

The authors are not aware of other KB attempts for usability assessment. The literature reveals the existence of KBS applications for damage evaluation of buildings. Most such attempts are based on evidence theory or fuzzy logic (eg., Ishizuka *et al.*, 1983; Ross *et al.*, 1990). The aims of these systems are different from those of Amadeus as they address the damage evaluation as such. In contrast, Amadeus handles this evaluation in the framework of the usability decision problem, which occurs in the context of the emergency situation after an earthquake. Also, the damage evaluation in Amadeus relies on visual inspection, and is based on rules of thumb developed by the expert. There also exist systems for the seismic vulnerability or risk assessment of building structures (Mijango, 1988; Miyasato *et al.*, 1986). However, these are different problems as they address seismic safety of existing buildings (with respect to future earthquakes) or seismic risk.

Justification of Expert Systems Approach

Why is knowledge-based technology suitable for this application? The basic criteria of desirability and feasibility of a KBS are satisfied in this case (Waterman, 1985). The knowledge involved is difficult to organize and formalize with conventional technology. Indeed, the conventional procedures involving questionnaires and manuals have been criticized in the past (Applied Technology Council, 1989a; Gavarini, 1985a). The recent methodologies, such as the ATC-20, have attempted to address the shortcomings identified. However, they still involve manuals and questionnaires. Even though the methods present a noticeable improvement over earlier ones, they are constrained because of the vehicle used for the knowledge transfer, namely the forms and manuals. Also, the real time interaction inspectors can have with a Knowledge-Based System cannot be equaled either in speed or in depth of reasoning with the use of guidelines in print. The interaction with a KBS is dynamic in contrast to the static interaction offered through manuals.

The ATC-20 provides a uniform set of definitions that all inspectors can use while performing the evaluation. It also identifies a set of situations that determine when a building is clearly unsafe by providing a check-list of conditions that are unacceptable for safety reasons. This ensures, to a reasonable extent, the same degree of conservatism, i.e., it allows for uniformity of decision. (Nonuniformity had been identified as a serious deficiency in past practices.) However, the reasoning involved in the decision making process has only one level of depth, and involves mostly structural damage surveying. In reality, the assessment requires a multi-disciplinary approach; it involves elements of seismology, geology, and structural and geotechnical engineering. It also necessitates a complex reasoning process.

However, the success of the ATC-20 methodology lies in its simplicity. The methodology was constrained to the format described earlier, focusing on a few key aspects of the problem. It did not involve the complex reasoning inherent to the problem so as to be practical and widely used. Indeed, the methodology can only be useful in this form since the vehicle used for transfer of information is static. Keeping in mind the emergency conditions in which the operation takes place, multi-level reasoning is difficult to implement so as to be clear and intelligible in questionnaires and manuals (Gavarini, 1985a; Pagnoni et al., 1989). When such methods are implemented, the resulting forms tend to appear cryptic and complicated to the user and hence loose their efficiency (Gavarini, 1985a).

Also, conventional computing techniques would be impractical as the domain is not well-defined. In addition, these techniques could not enable the flexible use of the domain knowledge, which is required in the task of assisting in a complex decision-making process. In fact, an attempt has been made to automate the decision process through traditional computing technology. A program was written in Basic to perform the evaluation, but this effort was not successful (Gavarini, 1987). There were no explanation facilities, because their coding required too much effort. In addition, debugging the program was an extremely time consuming task, and so were expanding, modifying and refining the knowledge. These problems were not characteristic of the chosen language of implementation, but of its underlying algorithmic nature. The knowledge-based technology was clearly superior to the algorithmic methods and quite suitable for this task which involves mainly heuristic knowledge and qualitative descriptions (attributes). Indeed, Knowledge-Based Systems can manipulate symbolic data effectively. They also enable manageable encoding of multi-level reasoning and flexible retrieval of the encoded information (powerful explanation facilities).

Furthermore, there exist recognized experts in the field. The expertise is scarce, valuable, and strongly needed in the event of a moderate or strong earthquake. The resulting KBS would have a high payoff considering the importance and the large scale of the problem. For example, in the Loma Prieta earthquake of October 17, 1989, while the experts agreed that the earthquake was not strong compared with what the area can experience, there were 27,000 buildings significantly damaged; of these, 1,400 have been demolished (UC Berkeley, 1990). Evidently, many more had to be routinely inspected. For the city of San Francisco alone, of the 8,000 buildings inspected, 1,200 were classified *with limited entry*, i.e., potentially unsafe, and 150, *unsafe* (Dames and Moore, 1989).

3. The Knowledge Acquisition Process

Planning for Knowledge Acquisition

Knowledge compilation and elicitation are best achieved through a review of the relevant documentation in the area, and direct interaction with domain experts. In this project, only one expert was involved. The usual difficulties associated with interacting with more than one expert, such as conflicts of opinions, were therefore not considered.

The problem at hand falls into the diagnostic category. The knowledge involved is fragmented and mainly heuristic. It is usually developed independently by experts through years of exposure to various postearthquake situations. The evaluations are solely qualitative, and represented by linguistic expressions. The language, which is consequently the principal vehicle of communication in this field, is fairly stable and well agreed on among experts for representing the domain and reasoning about it.

The expert for this project was articulate and enthusiastic about the system. He was therefore helpful and forthcoming with his explanations and comments. Furthermore, he was aware of the domain complexity and of the necessity of structuring and analyzing the knowledge involved. In addition, the knowledge engineers were relatively informed in the field, which made their interaction with the expert productive.

Knowledge Acquisition Techniques Used

Initially, the rapid prototyping approach was selected. This technique involves devising a simplistic model of the system to demonstrate some functionality and experiment with various approaches (M$_c$Graw and Harbison-Briggs, 1989). It is inherently iterative and requires frequent updating of the knowledge-base.

This technique was a means to get acquainted with the problem and experiment with different approaches to analyzing and representing it, mainly through trial and error. The principal sources of knowledge were documents provided by the expert. After studying the relevant literature, some informal and short discussions were held with the expert. These discussions helped provide a finer domain description. This phase led to establishing the functional objectives of the system.

The major phases in the KA process followed the typical framework of identification, conceptualization, formalization, implementation, and testing (M$_c$Graw and Harbison-Briggs, 1989). During the identification stage, knowledge engineers gain familiarization with the domain. In this case, this phase did not present any difficulty, and took relatively little time given that the knowledge engineers were well acquainted with the field.

Conceptualization requires the identification of the primary concepts involved in the domain and their relations with one another as described by the expert. This phase has been an ongoing process throughout the development of the project. Initially, the expert's prior work on the development of a postearthquake usability methodology helped considerably. Figure 1 shows the major factors influencing the usability decision. They are the geotechnical risk, the structural risk, and the complementary risk. This last element combines the effects due to external factors such as neighboring buildings, and those due to the nonstructural elements of the building itself. Further details can be found in Gavarini (1985a), and Pagnoni *et al.* (1989).

Formalization refers to the process of mapping all the relevant information conceptualized into formal representation mechanisms. This stage was carried out in conjunction with conceptualization for this project. Both phases are fundamental to the KA process as they involve constructing a pattern of both the expert's model of the domain and his solution generation schemes.

Implementation involves translating the formalized knowledge into the representational structure required by the selected expert system development tool. Here, this phase closely followed the formalization of the domain knowledge. Clearly all these stages are interdependent. This dependence is recommended, however, as long as it does not restrain considerably one aspect in favor of some other. Indeed, for a project to be successful, the formalization should neither completely ignore the tool of implementation, nor should it be forced by it. In this project, the tool selected for the final implementation had to meet a number of requirements. They were established by the knowledge engineers, after some preliminary development work was completed. An adequate representation of the domain knowledge was one of these conditions. The formalization accomplished for the final system represents the domain appropriately, and was approved by the domain expert, even though it was slightly guided by the selected development tool, PersonalConsultant[Plus].

Testing requires evaluating the resulting prototype for consistency and validity. Undeniably, it is important to test the system to verify that it is being built so as to meet the functional requirements set. Testing also measures the effectiveness of the various techniques used for knowledge representation and knowledge acquisition. It identifies problems related to the development itself. This phase is a cornerstone for incremental refinement and enhancement as it involves critically evaluating the evolving system. It usually leads to revisiting all previous stages in the search for more effective methods. In this project, it resulted in reconsidering the knowledge representation scheme and inference mechanism selected, as well as the conceptualization and formalization achieved for the development of the first rapid prototype. Thereafter, the knowledge engineers carried out ongoing testing of the rules for consistency and validity. Subsequently, the system was tested independently, in an on-site application (Gavarini, 1989).

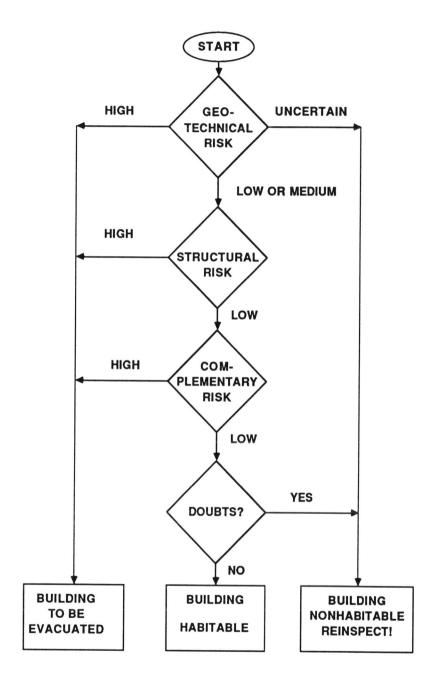

Figure 1: Flowchart of the Decision Process

After a knowledge representation scheme was agreed on, and a tool for the development of the final system was selected, KA continued in the form of incremental development. This well documented approach involves refining the system in increments of functional capability (M_cGraw and Harbison-Briggs, 1989). To allow for it, the system was structured to enable modification and expansion. These features are necessary to enable deepened representation of knowledge and enhanced reasoning capabilities. They are particularly important in this case as the field is relatively new and growing, and the knowledge involved is dynamic.

During the loops over the conceptualization-formalization-implementation-testing phases, a few structured interviews with the expert were necessary to answer some of the knowledge engineers' specific questions. Also, some additional informal discussion helped improve their understanding. These short and relatively frequent sessions were very helpful during the system's development stages. The opportunity to interact with the expert as the design of the system was evolving was extremely important and rewarding. Had the information obtained through these discussions been available at the beginning of the project, for example, it would not have been as useful to the project's development and completion. Indeed, it is the authors' opinion that when the knowledge query is gradual, the mental process the knowledge engineers go through during KA benefits their comprehension of the domain, and makes them more receptive to its subtleties. This process lets their understanding evolve and capture the concepts through a series of stages. The result is a sounder, better structured, and more refined knowledge-base.

Prototyping

As mentioned earlier, the rapid prototyping approach was chosen as a first stage of development of the Knowledge-Based System. The purpose was to test the feasibility of the project and to evaluate its practicality. It was also useful for setting the goals and functional requirements of the systems. It was agreed from the beginning to use an existing expert system development shell. The environment chosen was KEE[*] running solely on a Lisp machine, at the time. The developed system was to be used on-site, and therefore required a portable computer. At the time of the development of this first prototype, the authors were aware that another expert system development tool was needed for the final KBS. KEE was nonetheless initially chosen for its versatility and flexibility. In fact, Lisp machines serve as excellent prototyping workstations (M_cGraw and Harbison-Briggs, 1989).

---

[*]Knowledge Engineering Environment (KEE) is a frame-based software development environment with facilities for object oriented programming, rule-based reasoning, and data directed programming, developed by IntelliCorp.

The knowledge representation chosen for this first implementation was frame-based; it seemed well-suited to the problem. Representing the domain as objects was straightforward because there was an orderly classification that was easy to match. For example, the foundations of the building and its structural system can both be hierarchically classified under the object "building", as they belong to a building. Figure 2 shows the hierarchy of the frames in this representation.

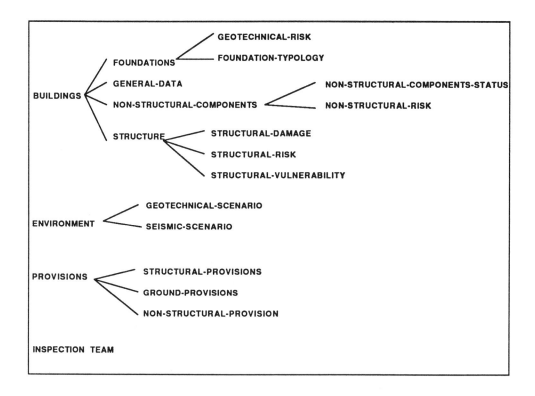

Figure 2: Hierarchical Organization of Frames in 1st Prototype

In this model, the domain is represented as a set of objects hierarchically classified in a tree-like structure. The classes are: (a) "Buildings", which contains all elements that are physically related to a building, such as foundations and structural system ("structure" in the figure); (b) "environment", which describes the geotechnical conditions and the seismic scenario for the specific site; (c) "provisions", which includes a summary of the practical measures to be taken, as suggested by the system (the application of these measures generally leads to rating the building as "habitable"); and (d) "inspection team", which contains all necessary information on the inspectors such as their professional background. The latter information is usually required by postearthquake survey forms. It is not involved in the decision process, and has been included here for completeness.

The end-element in each branch represents an attribute of the object at the preceding level, and is involved in the usability decision-making process. Some of these attributes are input-data such as "foundation-topology" and "seismic-scenario". Others, such as "structural-risk" and "structural-provisions", are to be determined in the course of the usability decision making; they are intermediate conclusions (or sub-goals) to be reached by the system.

The postearthquake questionnaire developed by the expert requires a large amount of data. Often, however, only a few data are needed to reach a usability decision. For example, if the observed damage strongly suggests that the structural system has seriously suffered from the earthquake, the building should be posted unsafe. It is not necessary, then, to evaluate the local geotechnical conditions. Nevertheless, it appeared that all the data required by the form are needed, and not only the elements that could lead to a usability decision. Indeed, gathering data on the earthquake and its effects appeared to be an essential part of the operation. Therefore, even though the problem at hand is diagnostic, a forward chaining (data driven) implementation was chosen. Consequently, the system would request sequentially all the data before reaching a final decision.

The rules act on the objects by altering the values of their attributes. In this formalization, there are two goal-levels. The geotechnical risk, the structural risk, and the complementary risk form the first one. Rules determine directly from the input data the values of these goals. The final decision, i.e., the usability decision, constitutes the second goal-level. The rules associated with this goal determine the final decision from the values obtained for the above risk quantities.

This first prototype demonstrated the feasibility of the project, but also highlighted a number of drawbacks in the approach taken. One important objective of the system was to help the inspector think about the situation by leading the reasoning, and explaining the various steps. Indeed, all conclusions have to be explained and justified. The forward chaining implementation showed weaknesses in this regard, as did the object representation and the rule design.

In the forward chaining design, all the data seem equally important to the inspectors. Indeed, the system asks the inspectors a series of undifferentiated questions whose answers are usually based on the inspectors' visual observations. Users are not lead through the reasoning process, as critical pieces of information are not emphasized. Therefore, they do not see which data drive the reasoning and are essential to the evaluation. Moreover, only at the end of the series of questions do the rules establish a value for the various risks involved, and subsequently, a value for the usability decision. How the usability decision depends on the risk values is quite clear from Figure 1. Undeniably, the determination of the geotechnical, structural, and complementary risks is the decisive evaluation the engineer must perform. It is the operation which incorporates all the major elements of the reasoning process. Unfortunately, the direct translation of the manual guidelines led to bulky rules. Consequently, they were not informative to the user due to the abundance of data in the rules' antecedents. Moreover, this practice resulted in difficulties related to the development; it was hard to keep track of the information while writing and implementing the rules because the conceptualization was too coarse. The explanation facilities of the system were working, and it was possible to request from it why a specific decision was reached. The system would then determine which rule(s) is responsible for that conclusion, and identify which value of the data led to it. However, the inspector would not gain any insight in the decision-making process through this practice as the rules would not specifically identify the concepts involved in the decision making.

A rule that embodies an important concept of the methodology is shown in Figure 3(a) to illustrate this point. It represents a set of conditions for which, even though the observed structural damage is apparently light, the resulting structural risk is high and the building is, consequently, to be evacuated. However, no clear concepts are identified in the given rule format. A mere reading of the rule does not explain the conclusion reached because the data are not grouped under characteristic headings. An inspector does not gain knowledge from this rule because no clear concepts are identified. Also the reading of the rule does not add to his or her understanding of the methodology. This rule will be reviewed in the next section in conjunction with the refining of the conceptualization. The purpose of this example is to illustrate the discussion. A detailed explanation of the elements involved is not warranted here, and can be found in Gavarini (1985a).

A more critical evaluation of the prototype, as well as the difficulties experienced while implementing the system, demonstrated that maintaining the knowledge base, validating it, and extending it was not a straightforward task. There were only two goal-levels, which rendered the reasoning shallow. Also, goals were associated with identified concepts. A limited number of goals reflected a limited number of identified concepts; therefore, it made the system lack explicitness. Likewise, most of the data were not grouped into meaningful concepts. Consequently, tracing the knowledge-base was an arduous task. It was evident that the system lacked modularity.

The conclusion was that the object representation and the forward chaining were unsuitable to this problem. Also, it appeared that the conceptualization had been done hastily and was therefore incomplete. However, this first attempt was considered rewarding because it proved the feasibility and usefulness of such a system. It also helped better understand what was needed, and what was to be avoided. It resulted in choosing backward chaining inferencing, because the associated explanations were judged more beneficial to the user. Indeed, as the implementation is goal-directed, the reasoning becomes more explicit to the inspector who then gains understanding of the process. To complete the form with information that may have not participated in the reasoning, but was nonetheless required, a series of "dummy goals" were created. They were designed so that completing the postearthquake form would not to interfere with the explanation capabilities of the system. Hence, this operation was designed to always take place

IF

           The Global-Structural-Damage-Level is LIGHT

      AND

           The Comparison-with-Historial-Max-Magnitude is
                    Less than max-magnitude minus 1

          OR

                    Not-known

      AND

           The Predicted-Sequence-of-Aftershocks is
                    More-Severe

          OR

                    Unknown

      AND

           The Building-Position is outside the epicentral area
           but inside the Seismic-zone affected by the earthquake

      AND

                  The Geotechnical Risk is MEDIUM

          OR

                  The Vulnerability is HIGH

THEN

      The Structural Risk is HIGH.

(a) - in Prototype

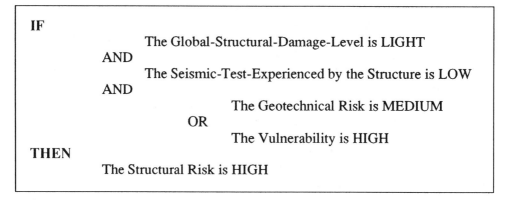

IF

           The Global-Structural-Damage-Level is LIGHT

      AND

           The Seismic-Test-Experienced by the Structure is LOW

      AND

                  The Geotechnical Risk is MEDIUM

          OR

                  The Vulnerability is HIGH

THEN

      The Structural Risk is HIGH

(b) - After Refined Conceptualization

Figure 3: Rule Examples

at the end to the session, after the the system reaches a usability decision. Also, PersonalConsultant[Plus] was selected as the tool of implementation as the features it offered seemed suitable. This point will be discussed more explicitly later. Finally, the knowledge representation scheme was altered as the conceptualization and formalization were becoming finer.

Conceptualization

The testing and evaluation of the prototype, together with additional interaction with the expert confirmed the need for further conceptualization. The need was for a crisper and more refined breakdown of the knowledge. The major factors affecting the usability decision had already been identified as the three risk concepts (see Figure 1). However, the concepts that lead to the determination of these factors in the methodology had not been clearly identified. The rules implemented did not isolate the few meaningful factors which affect the decision-making process, but enumerated as many factors as there were data. The task was not to identify new concepts that were not recognized earlier; it was to isolate from the knowledge-base elements that could be meaningfully grouped under one heading. These "new" concepts had always been present in the expert's mind, but were only taken into account implicitly in this first prototype. The task was to identify them.

The expert's work was reviewed and additional discussions were held with him. This made possible the breaking down of the evaluation of each risk concept into the evaluation of a number of sub-concepts; some of these, in turn, were further broken down. Doing this permitted the decomposition of the problem into a number of sub-problems each of which could be dealt with separately, in varying depth, depending on the state of the knowledge in the sub-area.

The determination of these sub-concepts was not straightforward. The concept of "conventional severity of the main shock" was explicitly identified by the expert. This factor combines the information on the magnitude and on the duration of the main shock of the earthquake. It is a qualitative, subjective measure of the severity of the seismic crisis in question. In some circumstances, it acts as a penalizing factor when the earthquake is judged severe (Gavarini, 1985a). However, most other concepts in the methodology were not isolated explicitly; the data they involved were not clearly specified. Nevertheless, the expert was talking quite comfortably about most concepts. It was therefore possible to isolate them through discussions. For instance, it appeared that the notion of "seismic test experienced by the structure" was an important element in the decision-making process. The expert explained how the seismic risk depended on the seismic-test; however, he did not state clearly what data elements were relevant to this concept. From the decision table he established for the evaluation of the structural risk, data were extracted and put under the heading "seismic test experienced by the structure" with the following possible values: extremely-strong, very-strong, moderate, light, and irrelevant (e.g. see Figure 3 and Pagnoni et al., 1989). The expert was satisfied with this classification, and with the elements involved in the evaluation of this concept.

The seismic-test combines information on the probable sequence of aftershocks, the earthquake magnitude and duration (its conventional severity), the relative position of the building examined with respect to the epicenter, and the magnitude of the maximum shock historically recorded in the area. This factor is a relative measure of the severity of the earthquake in relation to what can be expected in the near future (Pagnoni *et al.*,1989). Returning to Figure 3, with the newly identified concepts, the rule can be rewritten as shown in Figure 3(b). Now, a reading of the rewritten rule says the following: If the damage is light, and if the seismic-test is low, we need to look at the geotechnical risk and/or at the vulnerability of the structure. A medium geotechnical risk or a high vulnerability act as penalizing factors as the structural risk becomes high if either condition is satisfied. Here the concept of vulnerability identifies some types of construction and configurations that are known to offer poor earthquake resistance, such as unreinforced masonry construction (Gavarini, 1985a). With this reading of the rule, the inspectors understand the reasoning underlying the methodology. If they are confronted with a comparable situation, they know what information to look for, and what conclusions to reach. If, for example, the structural risk is light, the next step to take is to evaluate the "seismic test". The determination of the "seismic test" involves a few data which are compatible with its definition. Then, if the value determined is "LOW", one should check the geotechnical risk. (Note that if the geotechnical risk is high the building is classified as unsafe with no consideration of the other risks.) If the geotechnical risk is low, then one needs to evaluate the vulnerability of the building. Now the reasoning is consistent and meaningful. There is a template the inspectors can rely on in their evaluation.

Also, examining some of the rules that establish the value of the geotechnical risk, led to the identification of two separate concepts: the ground damage concept, which refers to the degree of damage to the ground, and the foundation-system-condition concept, which combines the structure's type of foundations, the topography of the particular location, and the type of soil. These two concepts were not clearly stated by the expert, however, he agreed with this breakdown. The background of the knowledge engineers helped considerably in the conceptualization stage.

The domain knowledge was partitioned through this process. The concepts identified are shown in Figure 4. They served as a basis to the new system. An intermediate goal parameter was associated with each concept. This is illustrated in Figure 5, where each box represents the sub-goal parameter associated with the identified concept. This way, the evaluation takes place step by step, as the various goals are evaluated in the process of making the usability decision. To keep the reasoning underlying the methodology clear and easy to follow, only a few data are necessary to the evaluation of each sub-goal and consequently of the concept associated with it. Reasoning on these concepts instead of reasoning directly on the input-data benefits the inspector's understanding and improves it. Indeed, it is easier to follow the stages of the decision-making process in the system on the identified concepts than on a multitude of input-data thrown together to give the final answer. Also the way the various sub-goals lead to the final decision is explained in the help facilities available in the system.

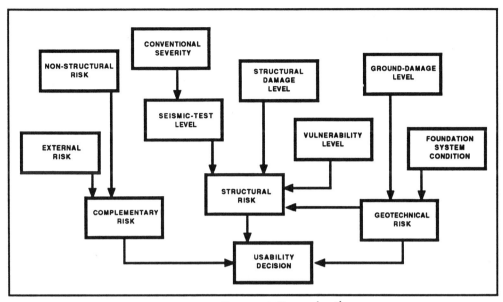

Figure 4: Concept Organization

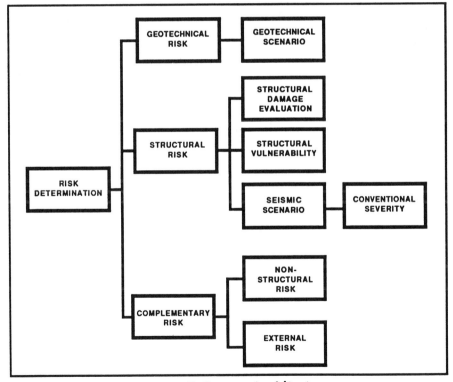

Figure 5: System Architecture

Furthermore, the modularity achieved served the development efforts well. It made both debugging and modifying the knowledge-base more tractable. Also, expanding and refining the KB can be easily managed as a result. This last feature has already been successfully tested. A module for more refined damage evaluation of masonry structures was added. A similar one is underway for reinforced concrete buildings.

Knowledge Representation

After the domain knowledge was coherently organized into interdependent concepts, the knowledge representation previously selected for the first prototype appeared unsuitable. Essentially, it did not take advantage of the increased modularity realized through the finer conceptualization. This led to considering and electing another KR scheme. It also resulted in the developing of a different system architecture.

The rule based representation appeared to be the most suitable because the knowledge is mainly heuristic, and in the form of IF-THEN rules. These rules present a traditional way to represent human knowledge, in particular to model human problem-solving activity, where such production systems have been successfully used (Duce and Ringland, 1988). They are particularly suitable in the case of diagnostic problems, which was the case in this project.

It appeared that representing the domain through the concepts that were identified through the improved conceptualization phase was a more meaningful approach than the previous one. In the first prototype the domain was divided into physical entities: a building, its foundations, the environment, the inspectors etc. In the new representation the domain objects are the concepts: the geotechnical risk, the structural risk, the seismic-test, the vulnerability etc. The concepts are organized hierarchically as shown in Figure 4. This representation differs considerably from the previous one where the few identified concepts were attributes of the domain objects. These domain objects mapped a physical breakdown of the domain. In contrast, the new representation was based on concept identification. Therefore, it inherently emphasizes the mechanisms of the methodology and makes the decision process readily available to the inspector. The basis for the representation shifted from a physical to a conceptual division of the domain. This change was an important result of the continuous knowledge acquisition effort.

The new representation proved to be valuable in making the reasoning-process understandable to the inspector. In fact, a major motivation in the development of a KBS for assisting inspectors in making usability decisions was to provide them with a clear methodology and extensive explanations. To illustrate the superiority of the new representation over the first one in this respect, consider what follows. Figure 4 readily informs the user that the usability decision rests on the evaluation of the geotechnical, structural, and complementary risks. It also shows that the evaluation of the structural risk depends on the level of structural damage, seismic-test level, structural vulnerability, and geotechnical risk. It also shows that the seismic-test depends on the conventional severity. The data that are involved in the evaluation of each concept have been left out of the figure, but are easily available through the system. The information obtained from Figure 2 which illustrates the previous architecture, does not help the understanding of the evaluation procedure as well. In addition, the seismic-test, which is an important concept in the methodology, is not at all present. Moreover, it could not be clearly identified as it would involve elements of the frame "building", and of the frame "environment". Similarly, the geotechnical risk would involve elements of these same frames. No gain in clarity can result from these observations.

In addition, the shell chosen made possible an almost exact mapping of the conceptual representation onto the system architecture (Figure 4 and Figure 5). The inference mechanism chosen was backward chaining because it was more suitable to explain the reasoning. Indeed, with the backward chaining, the system can answer questions such as why an information is necessary, and how conclusions are reached. (The shell used made these operations readily available.) Consequently, the reasoning is always focused on the goal and so is the inspector who, then, understands the steps he or she is taking toward reaching a usability decision. Finally, as mentioned previously, a series of dummy goals was introduced to allow the user to complete the questionnaire when necessary after the system reaches a usability decision. This operation does not interfere with the effectiveness of the explanation capabilities of the system.

Modality

A system's modality lies in its ability to communicate about the domain as well as its problem-solving capability (Breuker and Wielinga, 1987). It is a desirable quality as the knowledge accumulated in the system's knowledge-base can be valuable for a number of applications in addition to its problem-solving function. However, to achieve this multivalence, a KBS must be designed for such other uses.

A key element of the KA process lies in deciding how the user will utilize the system, and identifying the different classes of users (Kidd, 1987). In this project this aspect was of primary importance. It affected both the architecture of the system and the inference mechanism chosen. The modality of the system evolved along with the project. First, the emphasis was on the usability decision task. While the first prototype was being developed, what was expected from the system became clear.

Three roles were identified. First, the system should be a decision making aid, capable of reasoning and able to explain to the user the various stages of its reasoning. Second, it should also act as a data collection module; it should gather all the information on the earthquake and its effects, as well as some general data required by the questionnaire for subsequent data processing purposes. However, this function should not interfere with the system's ability to be transparent to the user, and to explain the reasoning throughout the evaluation. Finally, capitalizing on its explanation capabilities, the system should also assume a teaching function, and be used to train potential building inspectors for the usability decision task. The explanation facilities incorporated in the development shell facilitated this task. All three objectives were kept in mind while the final architecture for the system was being set.

Software and Hardware

The proposed system is to be used on-site, by building inspectors. It should therefore be running on a portable computer. The development shell chosen was PersonalConsultant[Plus] because it fulfilled the following basic requirements: it operates on a personal computer; also, a run-time version is provided for the developed system, and therefore, the memory requirements become minimal. They are easily met by a portable machine of the 286 class. In addition, the shell supports a rule-based representation, and both backward and forward chaining inferencing. It also permits a good representation of problems that can be broken down into a number of sub-problems that can, in turn, be broken down into more sub-problems etc. This decomposition allows a modular organization of pieces of information (data and goals) and ways to process this information (rules).

Furthermore, this shell facilitates the development phase. The available features permit tractable tracing of the knowledge-base. The shell allows grouping rules in various ways. For example, rules that result with the same value of the goal can belong to the same group. This greatly facilitates controlling the knowledge-base. Also comments can be attached to rules. They can briefly comment on the action of the various rules. In addition, the tool keeps an organized structure of the domain elements. For instance, a simple command lists the rules that affect any parameter value, as well as the rules where the parameter value affects a conclusion. Many more features facilitated the KA process.

The shell was derived from EMYCIN which is geared toward solving analysis (i.e., interpretation) problems (Van Melle, 1979). It uses a heuristic classification problem-solving method.

4. Conclusions

The KA process was a primary assignment in the development of Amadeus, a system for postearthquake usability evaluation. As the KA was evolving, so was the knowledge representation. In this experience, KA was an ongoing process, where, from very early in the process all stages were interacting continually. The conceptualization stage had a major impact in this project, and considerably affected the architecture and the meaningfulness of the KBS. The modularity that resulted from the conceptualization of the domain highlighted a few inconsistencies in the knowledge-base that would have otherwise been undetected. It also made the reasoning more transparent, and potentially more easily understood by the users. The KA techniques used were simple. They mainly consisted in reviewing the expert's writings on the topic, and carrying out some structured interview sessions with him in order to clarify certain situations. In this process the knowledge engineers' background in the domain area helped considerably. This is a desirable situation for achieving a successful KA phase and ultimately, a successful KBS.

Acknowledgment

The authors wish to thank Professor J. Connor, for his advice and encouragement throughout the project. Also, Zahra Tazir would like to express her gratitude to the Ministry of Higher Education of Algeria for sponsoring her work at MIT.

Appendix. References

Anagnostopoulos, S. A., Petrovski, J., and Bouwkamp, J.G. (1985). "Emergency Earthquake Damage and Usability Assessment of Buildings." *12th Regional Seminar on Earthquake Engineering,*, March 16-25, Halkidiki, Greece, 1-17

Applied Technology Council (1989a).*"ATC-20, Procedures for Postearthquake Safety Evaluation of Buildings.".*

Applied Technology Council (1989b).*"ATC-20-1, Field Manual - Postearthquake Safety Evaluation of Buildings.".*

Breuker, J., and Wielinga B. (1987). "Use of Models in the Interpretation of Verbal Data.", In *Knowledge Acquisition for Expert Systems: A Practical Handbook*, Alison L. Kidd (editor), Plenum Press, 17-24.

Dames, T. R., and Moore W. W. (1989). *"A Special Report on the October 17th, 1989, Loma Prieta Earthquake".*

Duce, D., and Ringland, G. (1988). *"Approaches to Knowledge Representation: An Introduction."* John Wiley & Sons Inc., 1-12.

Gavarini, C. (1985a). "A Proposal of an Inspection Form for Emergency Decisions on Buildings after an Earthquake." *Proceedings of a PRC-US-Japan Trilateral Symposium/Workshop on Engineering for Multiple Hazard Mitigation*, Beijing, China.

Gavarini, C. (1985b). "Agibilita' degli edifici dopo un terremoto: una proposta metodologica." *L'industria Italiana del Cemento*, 6.

Gavarini, C. (1987). *Private communication.*

Gavarini, C., Pagnoni, T., Tazir, Z. (1989) "Amadeus: un sistema esperto per la valutazione d'urgenza della agibilita' degli edifici dopo un terremoto." *L'industria Italiana del Cemento*, 1.

Ishizuka, M., Fu, K. S., Yao, J. T. P. (1983). "Rule-Based Damage Assessment System for Existing Structures." *SM Archives*. 8. Martinus Nijhoff Publishers, The Hague, Netherlands.

Kidd, A. L. (1987). "Knowledge Acquisition - An Introductory Framework.", In *Knowledge Acquisition for Expert Systems: A Practical Handbook*, Alison L. Kidd (editor), Plenum Press, 1-15.

M$_c$Graw, K. L., and Harbison-Briggs, K. (1989) *Knowledge Acquisition - Principles and Guidelines*, Prentice Hall Inc.

Mijango, C. (1988). "ASALVO: A Knowledge-Based Expert System for the Seismic Safety Assessment of Existing Structures," thesis presented to the Massachusetts Institute of Technology, at Cambridge, Massachusetts, in partial fulfillment of the requirement for the degree of Master of Science.

Miyasato, G. H., Dong, W., Levitt, R. E., Boissonnade A. C. (1986). Implementation of a Knowledge Based Seismic Risk Evaluation System on Microcomputers. *International Journal of Artificial Intelligence in Engineering*, 1, 1, 29-35.

Pagnoni, T., Tazir, Z., Gavarini, C., (1989). *"Amadeus: A KBS for the Assessment of Earthquake Damaged Buildings."* Expert Systems in Civil Engineering. IABSE Colloquium, Bergamo, Italy, 141-150.

PersonalConsultant[Plus], (1987). Reference Guide. Texas xInstruments Inc. Version 3.02.

Ross, T. J., Sorensen, H. C., Savage, S. J., Carson, J. M. (1990). DAPS: Expert System for Structural Damage Assessment. *Journal of Computing in Civil Engineering*, 4, 4, 327-348.

Van Melle, W. (1979). *"A Domain-Independent Production-Rule System for Consultation Programs."* IJCAI, 6, 923-925.

Waterman, D.A. (1985). *"A Guide to Expert Systems."* Addison Wesley.

CHAPTER 9

Inductive Learning of Bridge Design Knowledge

Yoram Reich
Steven J. Fenves

1. Introduction

Motivation

Knowledge acquisition for any expert system is time-consuming and tedious. This effort increases when dealing with design domains, which are ill-structured by their nature. One approach that promises to alleviate the difficulty of knowledge acquisition is the introduction of learning into the development and maintenance stages. Research in machine learning has produced a variety of learning algorithms that have the potential of acquiring expert knowledge. However, most of these algorithms were developed for and tested on classification or diagnosis tasks; their applicability to design domains has not yet been broadly proven.

The main focus of our research is to provide support in the preliminary design stage of large bridges. It is in this stage that most major design decisions and concept selections are made, leaving a tangible framework for detailed design. Our system is intended to aid bridge designers by *providing them with fast retrieval of potential conceptual solutions relevant to the design problem at hand.* The product of the research described here is not a self-contained design system but an aid that shortens the preliminary design stage. The system we are building can provide good candidate designs but does not guarantee their consistency or feasibility. Analysis and redesign of the candidates generated by our system requires additional machinery not addressed in the chapter. A parallel effort we are pursuing deals with these processes. The integration of these efforts will provide feedback on decisions made in the preliminary design stage. The framework will then allow for additional assimilation of knowledge from the system's experience.

Although cable-stayed bridges are the target domain, our initial experiments have used a simplified bridge design domain: the bridges of Pittsburgh. This domain contains descriptions of examples of 108 bridges constructed in Pittsburgh since 1818. Each description contains a set of specification attributes (parameters) and the description of the actual design in terms of a set of design description attributes. The task is to assimilate the examples and formulate knowledge that will help generate candidate design descriptions for given specifications of future bridges.

Terminology

This section defines basic terms used throughout this chapter.

169

Attribute. Each object is described by a list of attribute-value pairs. An attribute can assume several values if it has a discrete domain (e.g., the type of the bridge is cantilever), or infinitely many values if it has a continuous domain (e.g., the length of bridge is 1000 ft.). A major assumption is that in the domain of interest, objects generated in the preliminary design stage can be modeled by a list of attribute-value pairs.

Specification. The subset of attributes that represents the requirements for the object is called the specification (e.g., the bridge should have a vertical clearance of 70 ft.).

Design description. The subset of the attributes that represents the description of the physical properties of the object is called the design description (e.g., the main span of the bridge is 600 ft.).

Concept. An abstract description of a group of objects is called a concept. A concept should depict important attributes shared by the objects in the group.

Prediction. Prediction is the process that completes the description of a partially specified object. For example, if the description of a bridge contains only the specification attributes, prediction can be used to complete the description with the design description attributes. In this sense, the product of prediction is similar to the product of synthesis. In our approach, the prediction mechanism is entirely syntactic and domain independent.

Learning. A process that receives information about objects in a certain form (e.g., description of bridges) and transforms it into a different representation (e.g., abstract concepts) that can be used for prediction is henceforth called learning.

Knowledge Acquisition Approach

The project described started three years ago. The first step was to analyze existing machine learning techniques to assess their applicability and promise in design tasks (Reich and Fenves 1989b). We identified several categories of techniques suitable for different stages of the life cycle of an expert system. One of the promising paths of learning for knowledge intensive tasks such as design is the use of a general architecture with learning capabilities. We have explored such an architecture, Soar (Laird et al 1987), for a simple design task (Reich and Fenves 1988). Soar provides a mechanism for performance improvement by chunking successful solution paths into concise procedures; however, it does not have a mechanism for acquiring substantial amounts of knowledge from external sources.

For the task of knowledge acquisition we explored other machine learning techniques, starting with techniques that employ symbolic representations of concepts. The complexity of the learning task suggested that a single learning technique is not sufficient to produce the full functionality required. We identified the need to assemble and integrate several learning techniques. The integrated collection contains sufficient knowledge about learning techniques to allow for the selection of the most appropriate technique for different situations. The integrated collection constitutes a knowledge intensive learning system (i.e., it uses knowledge on how to use learning tasks); however, it is independent of any specific application domain. We have implemented a small part of such a system but have not tested it extensively (Reich and Fenves 1989a). The implemented part does not prove conclusively the applicability of the approach.

The difficulties manifested while working with symbolic representations of concepts suggested the exploration of techniques employing a probabilistic knowledge representation. This has led to the approach we are currently pursuing, which is appropriate for forming and using concepts.

Among the learning algorithms we have explored, COBWEB (Fisher 1987) seems to be well suited for organizing examples for future reuse, by extracting implicit knowledge from the examples and representing it in a hierarchical classification of concepts. COBWEB is incremental; it creates a hierarchical structure of concepts that supports top-down refinement; and it supports flexible prediction that can output several attributes based on the remaining attributes. Until recently, COBWEB had only been tested on classification tasks, where it has demonstrated very good performance.

We have developed Bridger, a system built on the foundations of the COBWEB algorithm, for learning in design domains (Reich 1991; Reich 1990a). Bridger extends COBWEB along several dimensions: it handles continuous attributes in addition to nominal attributes; it can forget irrelevant knowledge; it can correct the knowledge it learns; and it can use several prediction methods viewed as design methods. We have shown how COBWEB's prediction method is similar to case-based design, and how a new prediction method we have introduced can be viewed as top-down refinement. We have also established a mapping between Bridger and a general theory of design (Reich 1990a). Some of the extensions have been successfully tested in four design domains previously published in the literature (Reich 1991).

In this chapter, we describe Bridger's knowledge acquisition approach, using a case study of applying it to learn and use generalized concepts in the domain of Pittsburgh bridges. We extrapolate on the use of the approach to the more complex domain of preliminary design of cable-stayed bridges, which is the target domain of our system.

2. Background

In this section we elaborate on the target problem and on the domain in which we have experimented. We justify the use of the knowledge acquisition approach and outline its goals.

The Role of Preliminary Design

Fig. 1 provides an overall view of the process of bridge design (after G. F. Fox, (Spector and Gifford 1986)). A similar, less structured view is provided in (Leonhardt 1984). The design problem we are exploring is the enhanced preliminary design of large cable-stayed bridges, which is the first stage in the overall design process. Some aspects of the detailed design that may provide feedback to the first stage will also be incorporated, leading to the scope of work shown circled in Fig. 1.

Preliminary design of bridges is a complex task. Part of the difficulty arises from the fact that there are no good, generally applicable design procedures. This is exemplified by books on the design of cable-stayed bridges (Podolny and Scalzi 1986; Troitsky 1988). Most of these books contain a general discussion on possible configurations of bridges with some abstract guidelines for selection, mostly based on simplified case studies. This general discussion is then followed by the major part of the books, containing descriptions of actual bridges classified into several major sets.

Two criteria are used in the preliminary design stage for selecting between alternatives: an objective quantitative calculation of the cost and a subjective qualitative evaluation of additional considerations. The following example demonstrates this process. Fox describes

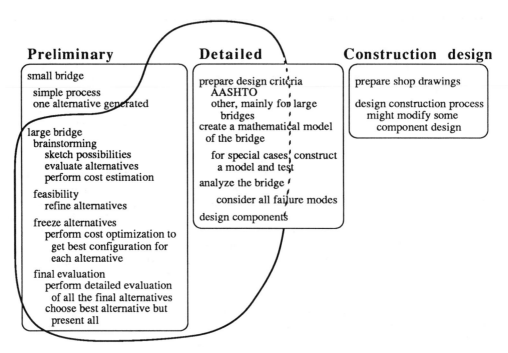

Figure 1. The Process of Bridge Design

the design alternatives for the Dame Point Bridge in Jacksonville, Florida (Fox 1979). The bridge spans a river 1800 ft. wide. Seven alternatives are provided with their relative evaluations. The cost, which is the objective criterion for evaluating the designs, is similar in all alternatives. Evaluations of esthetics, eased erection, interference with navigation during construction, etc. are given in the form of a comparison matrix. These evaluations are qualitative and unmeasurable. It is hard to infer what the best design is. A subjective judgment based on experience is used to select a partial set of alternatives for detailed consideration. The close competition suggests that: (1) for regular spans, several alternatives will always be available for consideration; and (2) the major decision is based on subjective criteria and experience, rather than on cost. In practice, expert designers are experienced with many past designs, allowing them to make the necessary subjective judgment.

Another part of bridge design books discusses analysis techniques. Analysis is an objective mechanism for evaluating designs, not for synthesizing them. Designing using objective judgment is hard since there are many inter-relations between parameters, so that their contribution to the final design cannot be isolated.

The last reason for the lack of a general design procedure results from the fact that cable-stayed bridges represents a rather new technology (only about 100 bridges constructed so far) and experience does not accumulate or dissipate rapidly.

Bridge designers can benefit substantially from a system that organizes past experience in a readily available format. Such a system can provide the most relevant information for the particular design scenario at hand. Furthermore, such a system can partially automate the process of preliminary bridge design. An important side effect of this work is the emergence of a better understanding of bridge design.

Remarks made by G. F. Fox, support the need for such an approach:

> "I think we should be developing databases, or knowledge bases, as well. Unit cost, technical data, historical costs, and failures could be put into these databases. Right now designers aren't able to get to the information that's out there (Spector and Gifford 1986)."
>
> "Failures usually occur when we extrapolate beyond our experience and models. From each failure, there's a lesson to be learned (Spector and Gifford 1986)."

There have been no substantial attempts so far to construct an expert system for this domain. There are some restricted studies that use optimization techniques to obtain a bridge configuration (Bhatti and Nasir Raza 1985). Certainly, no studies using knowledge acquisition techniques have been reported for bridge design, although related research in the use of learning in the design of wind bracing systems has been reported (Arciszewski and Ziarko 1986; Arciszewski and Ziarko 1988).

The lack of substantial previous work, the problem difficulty and its importance, and the existence of bridge examples described at various levels of detail in the literature make the domain amenable to a demonstration of the utility of learning in design.

Purpose of The Knowledge Acquisition Process

The purpose of the knowledge acquisition process is to learn concepts that can be used for making inferences of potential solutions based on the retrieval of successful designs or their combinations. The knowledge learned should not only facilitate inferences; it should do so in such a way that the complexity of the design process is reduced, by gradually allowing inferences that are more efficient than the initially available strategies, such as *generate and test*.

This process of constructing design descriptions from retrieved portions of past designs, in an efficient manner, will be referred to as 'designing.' It corresponds to the initial stage of design performed by expert designers (Leonhardt 1984).

Overall Approach

The development of a system that will improve its representation requires the implementation of a base-level design system (in our case, a bridge design system) and a learning program that will enhance the base-level system's performance. The base-level system can only rely on a simple, costly *generate and test* strategy.

Our approach to the use of learning in design is not restricted to knowledge acquisition, but also supports performance improvement over time. Fig. 2 provides an overall view of a system architecture based on our approach. The system consists of two main modules: the learning module and the redesign module. Examples of designs can be obtained from the literature or from designs produced by the system and evaluated by a human designer or a system critic. These examples are used to enhance the system's knowledge to design artifacts for given specifications. Candidate designs are given to a critic that evaluates them and submits them to a redesign system, if necessary (the redesign system is not necessary for the learning process, but it speeds up the process). Since the critic filters erroneous design solutions, the performance of the system need not be perfect; rather, we seek its improvement over time. The learning process makes use of the current state of knowledge and the new design example presented.

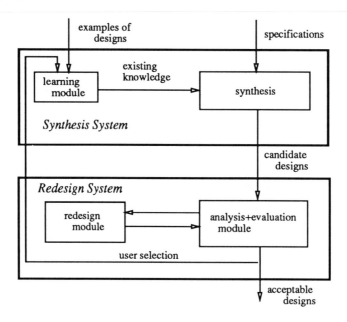

Figure 2. Incorporating Learning in Design

This approach to design is similar to the General Design Theory presented by Yoshikawa (Yoshikawa 1981; Tomiyama et al 1989), in which designs are represented as a list of attribute-value pairs, including attributes that describe the specifications and the designed artifact. The design process in the General Design Theory relies on some classification of artifacts into abstract concepts. However, when dealing with new a domain, there is no clear way of structuring it (at least not before learning takes place). Thus, it is not appropriate to impose a pre-defined design strategy such as top-down refinement; although, when appropriately constructed, it is a very efficient strategy. Also, there is no direct mapping between the goals of the design process and the functionality of available components; rather, many artifact attributes contribute to the achievement of the design objectives. The demand of not imposing a pre-defined strategy also applies to preliminary or conceptual design, where major design decisions need to be made without good quantitative measure of their tradeoffs. In particular, this demand is appropriate to our target domain of bridge design[1].

Every design domain is dynamic in nature. New technologies emerge, new materials are introduced, and domain theories are extended or replaced. The addition of a new attribute for describing designs can result in major problems for non-incremental learning systems after

---

[1]This view of design is different from common approaches to VLSI design, which use top-down refinement and constraint satisfaction and where design subgoals are matched against components or further decomposed (Steinberg 1987)), or to some similar mechanical design problems (Ramachandran et al 1988).

considerable knowledge has been accumulated. This dynamic nature necessitates the use of incremental learning in design systems. The incorporation of learning as an integral part of design systems must support two additional requirements. First, the learning system should handle complex description spaces that are appropriate for describing designs[2]. Second, the learning process should produce knowledge that operates in an acceptable design paradigm (for example, a top-down refinement strategy).

Domain of Pittsburgh Bridges

We formulated the following abstract preliminary design problem. The design process is initialized by specifying a set of *specification* attributes that the bridge has to satisfy (e.g., length of crossing, number of lanes or tracks, loading, horizontal and vertical clearance, etc.). Alternatively, only a partial set is provided, which can be viewed as an abstracted specification. The output of the preliminary design process is a set of attributes specifying the salient aspects of the selected design, termed the *description* attributes (e.g., material, type, key dimensions, etc.). The exact division into specification and design description attributes is not fixed. For example, a design description attribute can serve as a specification attribute if the client specifying the problem wishes to impose it as a constraint on the design. The design specifications and descriptions are described by lists of attribute-value pairs (e.g., lanes=4, type=cantilever, etc.).

The fundamental distinction between diagnosis and design is that in the former only *one* attribute value is to be inferred (i.e., the name or nature of the malady or dysfunction), while in the latter, *several* attribute-values collectively define the solution.

The domain on which the above model of design was tested is that of Pittsburgh bridges. The domain contains simplified descriptions of 108 bridges built in Pittsburgh since 1818. Each example is described by 12 attributes, 7 for the design specifications and 5 for the design description. The specification attributes describe the location of the bridge and the site characteristics:

- The IDENTIFIER and the NAME attributes are used to identify the examples and are not used in the classification process.
- The RIVER and LOCATION attributes specify the location of the bridge.
- The PERIOD attribute specifies the time the bridge was erected.
- The PURPOSE attribute specifies whether it is a highway or railway bridge.
- The LANES attribute describes the number of lanes for a highway bridge or tracks for a railway bridge.
- The LENGTH attribute is the total length of the crossing.
- The CLEAR-G attribute specifies whether a vertical navigation clearance requirement was enforced in the design or not.

The design description specifies the bridge configuration chosen to satisfy the requirements:

- The T-OR-D attribute specifies whether the bridge is a through or a deck bridge.
- The MATERIAL attribute specifies one of the following: wood, iron, or steel.
- The SPAN attribute is the length of the main span of the bridge.

---

[2]This does not contradict the representation as a list of attribute-value pairs, since attributes can be of several types: nominal (for example, describing the type of material), linear (discrete, for describing the number of spans, or continuous, for describing the length of the bridge), and hierarchical (describing various methods of realizing specific bridge types). Furthermore, there might be implicit relations between attributes.

- The REL-L attribute is the relative length of the main span to the total crossing length.
- The TYPE attribute describes the type of the bridge selected; which may be: (1) simple, continuous or cantilever truss; (2) arch; (3) suspension; or (4) wood.

Three examples of bridges are described in Table 1. Table 2 provides a summary of the attribute-values describing the example set. The top line in each sub-table provides the attribute names and the frequency with which they appear in the example descriptions. The following lines provide the attribute-values and their frequencies.

The attribute values provided in Tables 1 and 2 are all nominal. The discretization of the continues attributes was done in two stages. First, the domain of each continuous attribute was subdivided into several pre-defined ranges. Second, continuous values were assigned the name of the range that contained them. This discretization was performed for purposes of clarity. The experiments described in Section 3. use the original data.

Table 1. Examples of Bridge Descriptions

	attribute	Example 1	Example 2	Example 3
specification	NAME[1]	16th St. B.	Fort Duquesne B.	Penn. Turnpike B.
	IDENTIFIER[1]	E19	E84	E88
	RIVER	A	A	A
	LOCATION	29	24	37
	PERIOD	Craft	Modern	Modern
	PURPOSE	Highway	Highway	Highway
	LANES	2	6	4
	LENGTH	Medium	Short	Long
	CLEAR-G	N	G	N
description	T-OR-D	Through	Through	Deck
	MATERIAL	Wood	Steel	Steel
	SPAN	Medium	Medium	Long
	REL-L	Small	Full	Full
	TYPE	Wood	Arch	Continuous-Truss

[1]These attributes are not used in the learning process.

Limitations

At present, bridge descriptions (specifications and artifact description) must be described as lists of attribute-value pairs (continuous, ordered, or nominal attributes). This may seems a restriction; however, most description of bridges in books on bridge design are provided in this form (Leonhardt 1984; Podolny and Scalzi 1986; Troitsky 1988). To get better quality learning, we plan to extend the representation to deal with structured descriptions. In a structured description, a bridge is subdivided into a superstructure, a substructure and their assembly. Each of these parts is in turn decomposed into subparts that ultimately lead to elementary components described by lists of attribute-value pairs.

There are no other limitations to the approach. It can handle partially specified example descriptions, descriptions with irrelevant information, and descriptions with errors. Our experiments suggest that given enough experience, the relevant and correct knowledge will

Table 2. Summary of Attribute-Values

Specification

RIVER	108	LOCATION	107	PERIOD	108	PURPOSE	108
Y	3	Varies from 1 to 52		CRAFT	18	HIGHWAY	71
O	15			EMERGING	15	RR	32
M	41			MATURE	54	AQUEDUCT	4
A	49			MODERN	21	WALKING	1

LANES	92	LENGTH	81	CLEAR-G	106
1	4	SHORT	12	G	80
2	61	MEDIUM	48	N	26
4	23	LONG	21		
6	4				

Design Description

T-OR-D	102	MATERIAL	106	SPAN	92	R EL-L	103	TYPE	105
THROUGH	87	STEEL	79	LONG	30	FULL	58	CONTINUOUS-TRUSS	10
DECK	15	IRON	11	MEDIUM	53	MEDIUM	15	CANTILEVER-TRUSS	11
		WOOD	16	SHORT	9	SMALL	30	ARCH	13
								SIMPLE-TRUSS	44
								SUSPENSION	11
								WOOD	16

emerge from the examples.

3. Methodology of Knowledge Acquisition

This section describes the knowledge acquisition method. The processes of learning and design are specified and demonstrated on the domain of Pittsburgh bridges.

Requirements for Knowledge Acquisition

The only requirement for learning is a source of examples to be assimilated into the system's knowledge. Additional components can enhance the system's performance, as shown in Fig. 2. A critic that evaluates the system's output is important for filtering wrong designs. Filtered designs can be a source of examples to be avoided. A redesign module can complete the design process by providing solutions that pass the critic's judgment. Each design is the source of an example that should be considered in the future. The critic and redesign modules speed up the learning process by providing better examples to the learning system; however, they are not necessary for learning per se if examples can arrive from other sources. Consequently, these modules are not discussed further in this chapter. For a broader view of learning in design which contains these modules, see (Reich 1990b).

Knowledge Acquisition Process

Basic Process

Our model for the incorporation of learning in design is implemented in Bridger. Bridger is a domain-independent learning system for knowledge acquisition and performance improvement currently under development. Bridger is built on the foundations of the learning program COBWEB (Fisher 1987) and extends it along several dimensions. In particular,

COBWEB handles examples described by a list of *nominal* attribute-value pairs, whereas Bridger can handle entities described by a combination of nominal or continuous attribute types[3]. In addition, Bridger has a hierarchy-correcting module; it has a richer set of learning operators; it can forget undesired knowledge; and it can perform directed experimentation that increases the utility of its knowledge. Such extensions can improve the ability to master design domains. The following discussion concentrates on the knowledge organization and a simplified description of the performance of Bridger.

Bridger uses an incremental learning method for the creation of hierarchical classification trees. Bridger accepts a stream of designs described by a list of attribute-value pairs. Designs need not be classified as feasible, optimal, or by any other classification method. The classification emerges from the structure of the domain. Any *a priori* classification can be assigned to a design and treated as any other attribute. For example, the cost of an artifact may be used as a continuous evaluation attribute for classifying designs into "cheap" and "expensive" groups.

A classification is of a good quality if the description of a new design can be constructed with high accuracy, given that it belongs to a specific class. Such classification promotes inferencing, since it allows the prediction of attribute-value pairs based on class membership. Bridger makes use of a statistical function that produces a classification of a design set into mutually exclusive classes, $C_1, C_2, \ldots, C_n$ (originally proposed by Gluck and Corter 1985). The function used by Bridger is:

$$category\ utility = \frac{\sum_{k=1}^{n} P(C_k) \sum_i \sum_j P(A_i = V_{ij}|C_k)^2 - \sum_i \sum_j P(A_i = V_{ij})^2}{n} \qquad (1)$$

where :

C_k *is a class*,

$A_i = V_{ij}$ *is a attribute − value pair*,

$P(x)$ *is the probability of* x, *and*

n *is the number of classes*.

The first term in the numerator measures the expected number of correct attribute-value pairs that can be guessed correctly by using the classification. The second term measures the same quantity *without* using the classes. Thus, the category utility measures the *increase* of attribute-value pairs that can be guessed *above* the guess without the classification, i.e., based on frequency alone. The measurement is normalized with respect to the number of classes.

The term $P(A_i = V_{ij})$ is calculated by counters that store the number of times a design in the example set has the attribute-value pair $A_i = V_{ij}$ and the number of times a design has attribute A_i in its description. The term $P(A_i = V_{ij}|C_k)$ is calculated similarly by summing over designs in C_k only. $P(C_k)$ is the number of designs in class C_k relative to the total number of designs. This method has been extended to deal with continuous attributes in a similar manner (Reich 1991).

Bridger builds the classification hierarchy in the following way. When a new design is introduced, Bridger tries to accommodate it into the existing tree. Bridger starts its process from the root of the tree. The sub-classes of the root form the classification at this level of the tree. Given the new design and the current classification, Bridger can perform one of the

---

[3]CLASSIT, a variant of COBWEB accepts descriptions of continuous attributes only (Gennari et al 1989), but does not handle a combination of the two attribute types.

following operators:

(1) *expanding* the root, if it does not have any sub-class, by creating a new class and attaching the root and the new design as its sub-classes;

(2) *adding* the new design *as a new sub-class* of the root;

(3) *adding* the design *to one of the sub-classes* of the root;

(4) *merging* several sub-classes and putting the new design into the merged node; and

(5) *splitting* a sub-class and considering again all the alternatives[4].

Bridger uses the category utility function to determine the operator to apply: the best operator is the one that results in a new classification that maximizes the category utility function. If the design has been accommodated into an existing sub-class, the process recurses with this class as the root of a new tree. This control constitutes a simple hill-climbing strategy and results in the creation of a hierarchical structure of classes. Each class can be viewed as a generalization of all the classes below it.

Since the algorithm is incremental, there is no backtracking. Hence, the algorithm may suffer from unfavorable ordering of examples. However, the collection of operators (e.g., split and merge) allows for simulating backtracking, thus reducing the order effect (for additional details, see Fisher 1987). The incremental nature of learning is achieved by storing in each node statistical information about the designs stored below that node. This information is updated each time a new design passes through that node.

Design with Bridger

Bridger designs using a mechanism similar to learning, but allowing only operator 3, described above, to apply. Also, the statistical information stored at tree nodes is not updated. Bridger sorts the new specification through the tree to find the best host node in its hierarchy. The design progresses by assigning to the new artifact characteristic[5] attribute-value pairs describing the nodes traversed while sorting the specification through the tree. This strategy, which is different than the original method employed by COBWEB, can be viewed as a least-commitment process (Reich 1990a).

Fig. 3 shows the hierarchy generated by Bridger after learning 60 examples ordered by the date of their construction. The examples in this experiment are described by nominal attributes only. A node in the tree describes a generalized concept. The nodes in the hierarchy are described using their characteristic values only (shown in bold font). In addition, a distinct node name (e.g., G12 or G28) and the number of examples stored bellow a node appear above each node in the hierarchy. Vertical edges that do not lead to a node in the figure denote that the node below does not have any characteristics. The top class provides an overall summary of the example set. It summarizes the most frequent attribute-value pairs seen by Bridger. The nodes below the root of the tree differentiate the sample set into more refined classes.

As an illustration of the use of the hierarchy generated for design, we assume that the system is required to design a *highway* bridge on the *Allegheny river* where *vertical clearance governs*. Bridger starts designing by "classifying" the new design into the hierarchy. The

---

[4]Merging and splitting can be performed on combinations of nodes richer than is possible in COBWEB, although all the experiments reported later use the same operators as COBWEB.

[5]Characteristic are attribute-value pairs that are very frequent in all sons, i.e., $P(A_i = V_{ij}|C_k) \geq 0.75$ and that discriminate between classes, i.e., $P(C_k|A_i = V_{ij}) \geq 0.75$.

top node characterizes *through* bridges with 2 lanes. These attribute-values are assigned to the new design. Since vertical clearance governs the new design, the bridge description is refined using class G12, rather than class G28, where clearance does not govern. The new design will be of a *simple-truss* configuration made of *steel*. Class G19 is chosen next, since it represents *highway* bridges. Further refinement is done by using class G25 to decide on a *medium span* which is *full* relative to the length of the crossing. The design continues by using G27 to match the last specification, *Allegheny river*. The design is completed since all specifications have been met and all design attributes have been assigned. In cases where all specifications have been met but the design description is only partially completed, Bridger can use other refinement strategies to complete the design process, or to provide the partially described design as the solution (Reich 1990a).

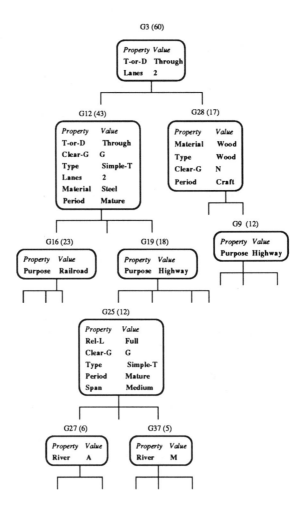

Figure 3. Design Concepts Hierarchy After Learning 60 Examples

Results and Verification

Evaluation Criteria

There are two measures for evaluating the system developed. The first measure concerns the performance of the system in the domain of application; it provides a purely quantitative evaluation. The second measure evaluates the quality or the 'naturalness' of the learning and design processes. These measures correspond to the following three criteria, variations of which are usually used for assessing learning systems:

- *Coverage*. How well can a subset of the examples reproduce the complete set? If a small subset suffices to reproduce the entire set, then the example set contains redundant examples. If, in addition, the example set is highly representative of the entire domain, we can say that the small subset covers the domain.
- *Performance*. How well does the knowledge extracted from the training examples transfer to unseen cases? This is the most important dimension for evaluating learning systems since the purpose of learning is to extract useful knowledge from examples and use it in the future.
- *Understandability*. How concise and understandable by humans is the learned representation of knowledge? This dimension is highly dependent on the representation language used. Learned knowledge should have an organization reflecting any intrinsic structure in the domain, and one should be able to understand the design generation process in terms of common design strategies.

The three criteria correspond to Michalski's dimensions of evaluating learning (Michalski 1986), namely, the *validity* (although coverage is more general), the *effectiveness*, and the *level of abstraction* of learning. Breiman et al (1984) gave the first two aspects precise statistical definitions as types of *internal estimates* for a classifier's error rate. Other quantitative dimensions for evaluating systems in general are the computational and memory costs of their operation and the sensitivity of their performance with respect to their design assumptions.

Quantitative Evaluations

Coverage is measured in the following experiment. After training, the examples are modified so that the learned knowledge can reconstruct them. This means that the design description attributes are erased, leaving the specifications only. The degree of fit between the reconstructed examples and the original examples is the validity. Measuring coverage means taking gradually increasing subsets of examples, training the system, and measuring the design performance on the complete set. After learning all the examples, coverage equals validity. Since the order of examples influences the quality of Bridger's learning, this experiment is performed n times with different random ordering of the examples. This experiment is called an *n-Random Coverage Experiment (n-RCE)*.

The second experiment measures the effectiveness of the approach by evaluating the transfer of knowledge to new problems. In this experiment, a subset of the examples is used for training and another, disjoint, subset for testing. Testing is done, as described before, by measuring the degree of match of actual examples with their reconstruction from partial information. This is a standard learning-performance test, but using *all* the unseen examples as test cases. Hence the number of test cases reduces as learning progresses. Such an experiment with n random orderings of the examples is called an *n-Random Performance Experiment (n-RPE)*. A better measure for large sets is the *V-fold Cross Validation (V-CV)* experiment. In this experiment, the example set is divided into V subset and V experiments are conducted where all subsets but one are used for training and the remaining subset for testing. The experiment is rerun with random divisions into subsets to collect statistics. Such an experiment is called an $n \times V$-CV, where n is the number of random divisions.

We conducted three experiments to assess the coverage and the performance of the approach. In measuring the coverage we used a 50-RCE experiment. In measuring the performance we used two experiments: a 50-RPE and a 10×10-CV. The results of the first experiment are described in Fig. 4a and those of the second in Fig. 4b. Results of the third experiment are summarized in the text and in Table 3.

In the figures, the horizontal axis describes the number of examples learned. It is important to note that the significant aspect is not the absolute number of examples, but rather the number of the examples learned relative to the total number of examples in the data set. In Fig. 4a, the vertical axis shows the accuracy of predicting the complete set of design attributes summarized over all the design attributes. In Fig. 4b, the vertical axis shows the accuracy of predicting the design attributes of the unseen examples based only on their specification attributes. This graph shows the accuracy for each attribute separately.

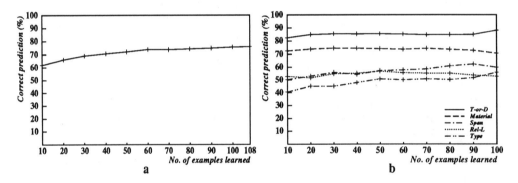

Figure 4. Design Performance: The Bridge Domain

Bridger improves its performance as more examples are assimilated. The coverage grows from 61% after learning 10 examples to 76%, suggesting that additional examples are required to achieve 100% coverage in this domain. Two attributes are easy to assign: the Material and the T-or-D attribute. The ability to predict them grows rapidly while learning the first 10 examples and then remains roughly constant. The prediction of the remaining three attributes improves slowly. Nevertheless, all these results are higher than by using the most frequent value for each attribute.

Performance in the 10×10-CV experiment, for both C4.5[6] (Quinlan 1989) and Bridger, is given in Table 3. In this experiment, there is a large variation between different runs, suggesting that more work should be directed at reducing the order effect in learning. Overall, Bridger's performance is slightly inferior to C4.5. One of the contributing factors for the difference is that Bridger constructs a single hierarchical classification whereas C4.5 constructs a separate decision tree for each design description attribute. Bridger can also construct a separate hierarchy for each attribute and improve its performance, but such hierarchies hide

---

[6]C4.5 is a descendant of ID3 (Quinlan 1986). It constructs decision trees for each design attribute treating it as a concept classification to be learned, assuming that the design description attributes are not correlated.

the real structure of the artifacts. We believe that in real, complex domains it is beneficial to learn and maintain a single hierarchy. In fact, a single hierarchy is sufficient to capture all the information needed; however, collecting this information requires using multiple paths when making predictions. Since efficiency is an important factor in making predictions, Bridger only uses a single path for its predictions. Note that better performance level will result when the examples reach 100% coverage and contain some redundant relevant information.

The results of the 10×10-CV and the 50-RPE experiments support each other. Whereas the first experiment is more statistically accurate, the second provides a learning graph and ultimately converges to the same results.

Table 3. Prediction Accuracy of Bridger and C4.5 on Pittsburgh Bridges Data

attribute	C4.5	Bridger			
	10-CV	10×10-CV		50-RPE	
		leaf[1]	normatives[2]	After 90 examples	After 10 examples
	Acc.	Acc.	Acc.	Acc.	Acc.
T-or-D	85	71	85	85	82
Material	85	77	75	73	72
Span	68	59	60	62	50
Rel-L	68	59	57	54	52
Type	56	47	54	52	40

[1] This is the original prediction method of COBWEB.
[2] This is the new prediction method of Bridger.

Since our target domain is more complex than the Pittsburgh bridge domain, we tested Bridger on an artificial domain with specific attributes. We constructed domains where the number of the following attributes can be varied: classes in the domain, attributes describing examples, specification attributes, design attributes, irrelevant attributes and noisy attributes. The probability of noisy data could also be changed. We experimented with several domains and concluded that the approach implemented in Bridger can handle these domains, and hence is expected to scale up to the cable-stayed bridge domain.

Other experiments tested successfully a hierarchy-correction technique, a forgetting mechanism, and an experimentation-directed knowledge acquisition strategy.

In the future we intend to construct other testing procedures that are more amenable to design domains. In these procedure a set of alternatives will be generated for each specification and compared to the original example. This reflects better the nature of design domains where many candidates are possible for most specifications. Such procedures will yield better results for Bridger.

Qualitative Evaluation

The results presented provide only a quantitative account of Bridger's performance. The first qualitative measure of Bridger's performance is the quality of the classification hierarchy it generates. We will examine the hierarchical organization of knowledge generated after learning 30, 60, and 108 examples of bridges, and analyze the transitions between them.

Fig. 5 provides the top layers of the classification tree generated by Bridger after learning 30 examples. The classes formed at the top levels are: a set of *through* bridges from the *emerging* period (node G12), a set of *highway* bridges (node G9), and an additional set of 4 bridges (node G7).

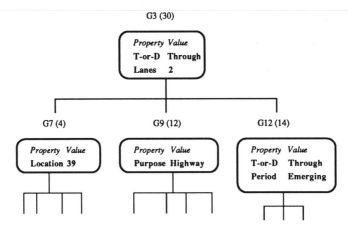

Figure 5. Knowledge Organization After Learning 30 Bridge Examples

Fig. 3, presented earlier, shows the tree after 30 additional examples have been learned. The two nodes of bridges from the *craftsman* period have been joined into one class with an additional *wood aqueduct* bridge to form a well defined class (node G28). This class is also characterized by the design specification of *clearance not governing*. The second class is comprised of *simple through truss* configuration of the *mature* period, made of *steel*, with *2 lanes* and with a design requirement of *clearance governing* (node G12).

This top level class configuration remains unchanged through the 100th example. All the examples are augmented to node G12 while node G28 remains static. After 5 more examples (101-105), node G12 breaks into two nodes (G16 and G61) using the split operator. The old node, G12, contains *steel*, *through* bridges with *clearance governing*; while the new nodes contain *railway* (node G16) or *highway* (node G61) bridges. Fig. 6 provides the final tree configuration after 108 examples. In the absence of characteristic values, dominant attribute-value pairs are displayed for several nodes (ordinary font below a horizontal line).

Comparing this tree with Fig. 3 reveals that several changes took place after learning the additional 48 examples. Since most of the recent bridges have more than 2 lanes, this attribute no longer characterizes the complete set. Bridger 'knows' about wider bridges and can use them when designing. The class of old bridges (G28) remains the same. It does not assimilate additional bridges. Class G12 (Fig. 3) was divided into two major classes: highway and railroad bridges, which are further decomposed. A careful look at Fig. 6 shows that class G24 consisting of two highway bridges is in a class characterized as being a railroad group. Two explanation exists for this apparent contradiction. The first explanation claims that this is an effect of the ordering of bridge examples. Such contradictions can be solved in Bridger by using the hierarchy-correction technique. The second explanation is that the two highway bridges are very similar to the class of railroad bridges in all other details. In addition, since statistical techniques drive this system, classes with so few examples do not constitute an important contradiction and do not change the overall behavior of the system.

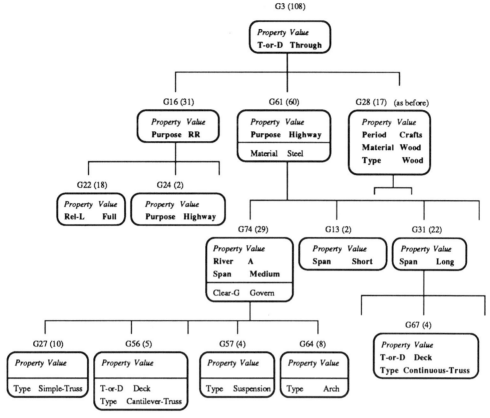

Figure 6. Final Knowledge Organization

An interesting hierarchy emerges below node G61. The sub-classes of this node are characterized by the *span* attribute. A design decision that is immediately influenced by a span choice – the *type* of the bridge – almost emerges as characterizing the classes at the lower level. This may be a sign that the hierarchy is capable of capturing the order dependency between design decisions.

The new knowledge structure would result in a different solution for the design problem previously illustrated. The design process would stop at node G74 resulting in a *through, steel* bridge with a *medium* span. This is an abstract description of a bridge since the type of bridge and its relative span length are not specified. Due to the additional knowledge, Bridger can now design at a higher level of abstraction to satisfy abstract specifications. In the previous design scenario, too many commitments have been made due to small experience.

The classification hierarchies described show that Bridger converges rapidly to a stable tree that depicts the important features of the bridge classes. Bridger modifies the hierarchy only if sufficient additional information suggests that it will benefit the inference ability.

The second qualitative evaluation is the interpretation of Bridger's design process in terms of common design strategies. The process of design, as illustrated before, corresponds to a top-down refinement with a least commitment strategy. Other design procedures available in Bridger, such as case-based design, are described in Reich (1990a).

Intended Mode of Operation

As illustrated in Fig. 2, Bridger is envisioned to operate in the following manner in a new domain. Initially, it will be supplied with a set of attributes and their possible types (continuous, discrete, etc.) and values. Bridger will be augmented with a critic (heuristic or analytical, such as an analysis program) that can evaluate designs. Optionally, a collection of successful designs will be provided. In their absence, Bridger will generate designs and evaluate them using its critic. Acceptable designs will be learned by Bridger that will gradually focus its strategy from a random generation process to a guided construction from successful designs stored in the hierarchy. As time passes, the number of generations before an acceptable design is achieved for a given specification will decrease, since the critic's knowledge will be augmented to the hierarchy. When Bridger is not used by a human designer, it will generate problems that increase its ability to design, solve, evaluate, and learn from them.

During Bridger's operation, the critic may be replaced and new attributes or values might be added. This will locally slow the learning process, but Bridger will overcome the shift in the domain.

In the limit, Bridger's performance will converge to a highly skilled behavior that will design adequate bridges by construction from previous successful designs. Analysis or redesign procedures will be used only for final verification of the design and for performing small modification needed.

4. Conclusions

Evaluation

The project is currently at its half stage. The overall approach has been proved for simplified design domains (Pittsburgh bridges and others, see Reich 1990a). Additional tests on benchmark learning databases have proved that Bridger can predict continuous attribute values as well as discrete values. Several simple tests on artificial domains support the scaleability of the approach to more complex design domains. We have also obtained some theoretical support for the approach.

The learning technique used gives rise to design procedures that match our understanding of design processes: top-down refinement, case-based reasoning, etc. In addition, it generates an organization of generalized concepts that is relevant to the domain.

Experience Gained

Overall, the study conducted increases our confidence in the potential of automatic knowledge acquisition for design domains.

The success of the approach to date relies on a careful match between design descriptions available in the literature and the required representation of examples in the knowledge acquisition approach developed. In the bridge domain, a description of a preliminary bridge design can be naturally represented by a list of attribute-value pairs, and the learning algorithm presented operates well on such representations.

Our experience to date indicates that probabilistic representations are more appropriate for the first stages of knowledge acquisition in design domains than symbolic representations. This is due to the ill-structured nature of design domains. In addition, we experienced that purely syntactic learning techniques can expose a considerable amount of the semantics of design decisions and can represent it as a simple mapping. Such techniques have a potential of being useful in applications other than design when organization and intelligent indexing of information is important.

We also performed restricted experiments with augmenting the syntactic learning techniques with domain knowledge. The enhanced performance obtained suggests that additional incorporation of knowledge may benefit the learning performance.

Recommendations

A definite recommendation from our studies to date is that the types of knowledge acquisition tools examined need further development and testing. We intend to complete the work on cable-stayed bridges. Successful performance in this domain will strengthen the support for our approach.

Further enhancements of the approach can be obtained by providing a more interactive environment between the designer and the learning system (e.g., open the learning and design processes to allow the expert to override decisions).

This study illustrates that learning requires large databases of information, whose collection and organization is a time-consuming process. Nevertheless, such databases can ease the process of design. The development of databases for future use in learning, case-based design, etc. will pay off in better future performance.

We hypothesize that the technique presented can be successfully used to organize data in other domains if pieces of data can be assigned indexing information in the form of a list of attribute-value pairs.

Acknowledgments

This work has supported in part by the Engineering Design Research Center, a National Science Foundation Engineering Research Center, and the Sun Company Grant for Engineering Design Research. We would like to thank Ross Quinlan for providing the results of running C4.5 on the Pittsburgh bridges database and to the reviewers for their constructive comments.

Appendix. References

Arciszewski, T. and Ziarko, W. (1986). "Adaptive expert system for preliminary engineering design." In *Proceedings of the Sixth International Workshop on Expert Systems*, pages 695–712, Avignon, France.

Arciszewski, T. and Ziarko, W. (1988). "Adaptive expert system for preliminary design of wind bracing in steel skeleton structures." In Beedle, L. S., editor, *Second Century of the Skyscraper*, pages 847–855, Van Nostrand Reinhold, New York.

Bhatti, M. A. and Nasir Raza, S. M. (1985). "Preliminary optimal design of cable-stayed bridges." *Engineering Optimization*, 8(4):265–289.

Breiman, L., Friedman, J. H., Olshen, R. A., and Stone, C. J. (1984). *Classification and Regression Trees*, Belmont, Waldsworth, CA.

Fisher, D. H. (1987). "Knowledge acquisition via incremental conceptual clustering." *Machine Learning*, 2(7):139–172.

Fox, G. F. (1979). "The Dame Point Bridge main spans-superstructures." In *Long Span Suspension Bridges: History and Performance*, American Society of Civil Engineers, New York, preprint 3590 edition.

Gennari, J. H., Langley, P., and Fisher, D. (1989). "Models of incremental concept formation." *Artificial Intelligence*, 40(1-3):11–61.

Gluck, M. and Corter, J. (1985). "Information, uncertainty, and the utility of categories." In *Proceedings of the Seventh Annual Conference of the Cognitive Science Society, Irvine, CA*, pages 283–287, San Mateo, CA, Academic Press.

Laird, J. E., Newell, A., and Rosenbloom, P. S. (1987). "Soar: an architecture for general intelligence." *Artificial Intelligence*, 33(1):1–64.

Leonhardt, F. (1984). *Bridges Aesthetics and Design*, MIT Press, Cambridge, MA.

Michalski, R. S. (1986). "Understanding the nature of learning: issues and research directions." In Michalski, R. S., Carbonell, J. G., and Mitchell, T. M., editors, *Machine Learning: An Artificial Intelligence Approach, Vol 2*, pages 3–41, Tioga Press, Palo Alto, CA.

Podolny, W. and Scalzi, J. B. (1986). *Construction and Design of Cable-Stayed Bridges. Second edition*, John Wiley and Sons, New York.

Quinlan, J. R. (1986). "Induction of decision trees." *Machine Learning*, 1(1):81–106.

Quinlan, J. R. (1989). "personal communication.".

Ramachandran, N., Shah, A., and Langrana, N. A. (1988). "Expert system approach in design of mechanical components." *Engineering with Computers*, 4(4):185–195.

Reich, Y. and Fenves, S. J. (1988). "Floor system design in soar: A case study of learning to learn." Technical Report EDRC-12-26-88, Engineering Design Research Center, Carnegie Mellon University, Pittsburgh, PA.

Reich, Y. and Fenves, S. J. (1989). "Integration of generic learning tasks." Technical Report EDRC 12-28-89, Engineering Design Research Center, Carnegie Mellon University, Pittsburgh, PA.

Reich, Y. and Fenves, S. J. (1989). "The potential of machine learning techniques for expert systems." *Artificial Intelligence for Engineering Design, Analysis, and Manufacturing*, 3(3):175–193.

Reich, Y. (1990). "Converging to "Ideal" design knowledge by learning." In Fitzhorn, P. A., editor, *Proceedings of The First International Workshop on Formal Methods in Engineering Design*, pages 330–349, Colorado Springs, Colorado.

Reich, Y. (1990). "Design knowledge acquisition: Task analysis and a partial implementation." *Knowledge Acquisition*, in Press.

Reich, Y. (1991). "Building and improving design systems: A machine learning approach." PhD thesis, Department of Civil Engineering, Carnegie Mellon University, Pittsburgh, PA.

Spector, A. and Gifford, D. (1986). "A computer science perspective of bridge design." *Communications of the ACM*, 29(4):268–283.

Steinberg, L. I. (1987). "Design as refinement plus constraint propagation: The VEXED experience." In *Proceedings of AAAI-87, Seattle, WA*, pages 830–835, San Mateo, CA, Morgan Kaufmann.

Tomiyama, T., Kiriyama, T., Takeda, H., Xue, D., and Yoshikawa, H. (1989). "Metamodel: A key to intelligent CAD systems." *Research in Engineering Design*, 1(1):19–34.

Troitsky, M. S. (1988). *Cable-Stayed Bridges: An Approach to Modern Bridge Design. Second edition*, Van Nostrand Reinhold, New York.

Yoshikawa, H. (1981). "General Design Theory and a CAD system." In Sata, T. and Warman, E., editors, *Man-Machine Communication In CAD/CAM, Proceedings of the IFIP WG5.2-5.3 Working Conference*, pages 35–57, North-Holland, Amsterdam.

CHAPTER 10

Inductive Learning of Wind Bracing Design for Tall Buildings

Mohamad Mustafa
Tomasz Arciszewski

1. Introduction

One of the most difficult and time consuming tasks in the design of tall buildings is the design of their wind bracings. The structural behavior of individual types of wind bracings is relatively well understood and the available analytical tools are usually sufficient for practical purposes. The design of wind bracings is typically conducted as a three-stage process. In the first stage, a class of wind bracings to be considered is selected, for example, rigid frames or truss bracings. Next, a single type, or several types of wind bracings are chosen. Finally, the complete analysis, design and optimization of individual wind bracing types are conducted. One of the main problems is the selection of the most appropriate type of wind bracing for a given building, considering many selection criteria, such as function, stiffness characteristics, weight, cost, fabrication, etc. Because of the complexity of the problem, it is difficult to use any formal multicriteria evaluation and selection models to select an optimal type. Therefore the wind bracing type is usually selected through an informal analysis conducted by experienced designers. This analysis usually results in one, two, or several types of wind bracings which are considered feasible and appropriate, but not necessarily optimal, in a given building. The practice described has several disadvantages: it requires the involvement of experienced designers, it is very time-consuming, and it usually produces feasible but not optimal results.

At present, the analysis, design, and optimization of wind bracings in steel skeleton structures can be conducted using computer packages such as SODA[1]. The use of these packages makes the design of wind bracings easier, but it still does not eliminate the need for the proper selection of a wind bracing type for a given design case. This selection has to be based, as before, on the designer's experience. A computer package can produce an optimal design of any type of wind bracing, including types which, although feasible, may not be the best in a given case. For example, a computer package may produce a minimum- weight design for a wind bracing in the form of a one-bay rigid frame in a twenty story skeleton structure. This design may not be a global optimal design: much better weight and stiffness characteristics might be obtained with a truss

---

1 SODA is a Structural Optimization, Design and Analysis computer package for planar steel frames and trusses under static loads. It has been developed by Waterloo Engineering Software, Ontario, Canada.

wind bracing. The problem of the selection of the proper type of wind bracing is particularly important when inexperienced designers use design and optimization computer packages; their lack of experience may lead to feasible designs which could be significantly improved through simple changes in configuration.

There are knowledge-based systems for the selection of types of wind bracings (Maher 1984 and Sriram 1987). These systems were prepared chiefly for experimental purposes and to demonstrate the feasibility of developing such systems for structural design. The knowledge used in these systems was manually acquired and was sufficient for demonstration purposes. The systems were never intended for practical applications, and therefore knowledge acquisition and its completeness were not a problem. Our intention has been to develop a practical tool for the selection of wind bracings, and therefore knowledge acquisition is important. Because we were aware of the problems associated with manual knowledge acquisition in such a complex domain as tall buildings, we decided to use automated knowledge acquisition.

The objective of our knowledge acquisition project was to acquire knowledge in the form of decision rules governing the design of wind bracings in steel skeleton structures. These decision rules should identify the relationships between the individual components of a wind bracing and its recommended height. The desired knowledge should be suitable for a knowledge-based decision support system for assisting structural designers in the selection of wind bracing types.

2. Planing for Knowledge Acquisition Process

In the first stage of our project, an engineering methodology of inductive learning was developed (Arciszewski et al 1987, Arciszewski and Mustafa 1989, Mustafa and Arciszewski 1989). It includes multistage strategies for conducting an inductive learning process, a method of example selection for use in subsequent stages of the learning process, and a method of monitoring the learning process using the proposed system of learning control parameters of both global and local character (Arciszewski and Mustafa 1989).

The second stage included the selection of a formal taxonomy of wind bracings and the determination of recommended heights for individual types. We decided to use the taxonomy of wind bracings proposed in Arciszewski (1985). This taxonomy is based on a ten-attribute description of a wind bracing. Nine attributes describe the individual structural components, while the last attribute "J. Wind Bracing Family of Types" identifies the family of types to which a given type belongs; (e.g., frame bracings, truss bracings, etc.) The attributes include nominal characteristics with symbolic values, such as the attribute "A. Static character of Joints," and numerical attributes of discrete character, such as the attribute "B. Number of Bays Entirely Occupied by Bracing." A complete description of a particular wind bracing is identified by a unique combination of values of attributes A through I. The attributes describing a wind bracing are as follows:

A. Static character of joints
B. Number of bays entirely occupied by bracing
C. Number of vertical trusses
D. Number of horizontal trusses
E. Number of horizontal truss systems

F. Material used for core
G. Number of cores
H. Structural character of external elements
I. Static character of external bottom joints
J. Wind bracing family of types.

The attributes used are based on the following definitions of the structural components of a wind bracing (Arciszewski 1985):

A vertical truss is a truss with parallel chords situated vertically in the main plane bracing.

A horizontal truss is a truss with parallel chords situated horizontally in the main plane bracing or in the plane parallel to it.

A horizontal truss system is a spatial truss grid composed of horizontal and transverse horizontal trusses.

A core is a system of interacting frames, trusses, frame walls, or reinforced concrete walls, located in different planes.

External elements are structural members of bracing situated in the plane of the external walls.

External joints are structural joints of external members. The bottom external joints are the external joints between external members and the foundation.

Two additional attributes were added to the existing description. These attributes identify the recommended upper and lower height ranges for individual types of wind bracings:

K. Minimum recommended height range
L. Maximum recommended height range

The taxonomy of wind bracings used in our project was initially developed to classify individual wind bracing types, identified by feasible combinations of attributes A through I, to one of the wind bracing classes, called "J. family of types". In this case all ten attributes A through J had to be used including the wind bracing family of types attribute. The objective of the project reported here was knowledge acquisition, and we were interested determining of the relationships between individual structural components of a wind bracing, or their combinations, and the recommended wind bracing height. Therefore, the wind bracing family of types attribute was not included in the examples used in our analysis. All the attributes used in our project and their possible values are summarized in Table 1.

These attributes were used to prepare 164 examples. Each example was for a different type of wind bracing, identified by the combination of attribute values, and included minimum and maximum recommended heights for this type. The types belonged to all four families of wind bracings as specified in Arciszewski (1985).

ATTRIBUTES	ATTRIBUTE VALUES						
	1	2	3	4	5	6	7
A. Static character of joints	Rigid	Hinged	Rigid & Hinged				
B. Number of bays entirely occupied by bracing	1	2	3				
C. Number of vertical trusses	0	1	2	3			
D. Number of horizontal trusses	0	1	2	3			
E. Number of horizontal truss systems	0	1	2	3			
F. Material used for core	0	Steel	Reinf. Concrete	3			
G. Number of cores	0	1	2				
H. Structural character of external elements	0	Columns & Beams	Cables	Bar Discs			
I. Static character of bottom external joints	0	Rigid	Hinged	Rigid & Hinged			
J. Wind bracing family of types	Frame	Truss	Frame - Truss	Core			
K. Minimum recommentded height	< 10	10 - 16	16 - 24	24 - 32	32 - 40	40 - 60	> 60
L. Maximum recommended height	< 10	10 - 16	16 - 24	24 - 32	32 - 40	40 - 60	> 60

Table 1. Taxonomy of Wind Bracings and their Recommended Heights

Our examples were prepared by consulting researchers and structural designers specializing in the analysis and design of tall buildings, from knowledge available in various books, and through published and unpublished results of comparison studies conducted on truss wind bracings (Schueller 1987, Scalzi 1981, Taranath 1988, Arciszewski et al 1988). One example from our collection will be discussed here to explain the structural interpretation of individual attributes used in this project. This example is:

A = 1, B = 1, C = 1, D = 1, E = 1, F = 1, G = 1, H = 1, I = 1, and K = 1 (less than 10)

The character of joints is described by attributes A and I. For the bracing under discussion, the attribute A = 1, indicates rigid joints, and the attribute I = 1, indicates no external joints since the bracing is located in the central bay of a three-bay skeleton structure. This is also inferred by attribute B = 1, i.e., the number of bays occupied by the bracing is one. The bracing is in the form of a simple moment-resisting frame. It does not have any core, vertical trusses, horizontal trusses, or systems of horizontal trusses. Therefore attributes C, D, E, F, G, and H are assigned a 1, indicating that no structural component is defined by these attributes. A single-bay moment-resisting frame is usually recommended by experts for a building of less than ten stories, and for this reason the value of attribute K equals 1, i.e., the minimum recommended height range is under ten stories.

In the third stage of our project several inductive systems were compared and considered for further use, including Super-Expert, BEAGLE, and ROUGH. These systems are briefly described in Chapter 3 of this monograph, "Machine Learning in

Knowledge Acquisition." BEAGLE was eliminated for two reasons: (1) it was developed for analyzing examples with continuous attributes rather than categorical ones and (2) its generation of decision rules is based on Darwinian evolution, and is random, and thus its results are unrepeatable and could not be compared with other systems under various learning strategies. Super-Expert was eliminated because it cannot produce rules from contradictory examples, which are usually quite prevalent in complex domains. Therefore we decided to use ROUGH in our project, particularly since we had gained some experience using this system earlier and received good technical support by the developers of this package, Voytech Systems, Inc. (Arciszewski and Ziarko 1987, 1988 & 1990).

3. Implementing the Knowledge Acquisition Process

The objective of our automated knowledge acquisition process was to determine the relationships between the individual components of a wind bracing and its recommended height. Therefore attributes A through I were considered as independent attributes while the recommended heights, attributes K and L were assumed to be dependent.

Several different strategies of inductive learning were used. These strategies were developed as a part of the engineering methodology of automated knowledge acquisition prepared in the initial stage of our project and reported in (Arciszewski and Mustafa 1989). All these strategies are multistage. First, the entire body of examples is divided into several groups, and the inductive system is used on the first group of examples, then the second group is added, inductive learning is repeated, etc. The examples used and the results obtained at individual stages of the learning process are recorded and monitored to determine the progress of learning and to establish when it is completed.

Five basic strategies of inductive learning were used. In the Point Strategy all available examples were used in a single learning stage. The Uniform Linear Strategy used at least two learning stages. The set of all available examples was divided into at least two subsets with equal numbers of examples. The first subset was analyzed in the first stage of the learning process, then the second subset was added in the second stage, and the learning was repeated.

There were three types of Mixed Point strategies used, in which all examples are divided into sets of significantly different sizes. In the Front-Point Mixed Strategy the largest set of examples was used in the first stage, while in the Central-Point Mixed Strategy it was added at the central stage of the learning process. In the Back-Point Mixed Strategy the largest set of examples was added at the last stage.

In general, there are four strategies of example selection and sequencing that can be used with learning strategies: Balanced, Intuitive, Purely Random and Corrective Strategies. In the Balanced Strategy, all available examples are analyzed, and carefully divided into individual sets by one or more domain experts. Each set is supposed to represent the entire spectrum of cases covered by the entire collection of examples. In the Intuitive Strategy an expert selects examples for individual stages of the learning process without analyzing them. This selection is called intuitive, because the expert's selection is not purely random and he/she makes subconscious choices derived from their experience.

In the Purely Random Strategy all examples are numbered and their sequence is randomly determined using a computer program to select examples for individual sets. The Corrective Strategy is intended to correct wrong decision rules, as they are noted by a domain expert monitoring the learning process. In this strategy an additional group of examples is prepared to cover aspects of the domain knowledge which were inadequately covered by the main body of examples and which led to wrong decision rules requiring correction. The use of these additional examples should improve the decision rules and eliminate wrong ones. In our project only the Purely Random and Intuitive Strategies of example selection in conjunction with the five strategies of inductive learning were used.

The automated knowledge acquisition processes were monitored to evaluate the learning process and to determine when learning was completed. This was done using three process control parameters, selected from a larger class of such parameters proposed in Arciszewski and Mustafa (1989). The knowledge acquisition process was monitored by constructing learning curves for individual control parameters and by building decision rule networks.

The three control parameters were as follows. The Global Rule Control Parameter measured the number of decision rules developed at individual stages of the learning process. The Global Attribute Control Parameter measured the number of attributes used in developing decision rules, again at individual stages of the learning process. The Local Attribute Control Parameter was the minimum number of attributes changed at individual stages of the learning process, including those added or deleted.

A learning curve is a graphical representation of a relationship between a given learning control parameter and the number of examples used at individual stages of the learning process. Examples of learning curves for the three control parameters are given in Fig. 1. These curves are for decision rules related to the conclusion "minimum recommended height range less than ten stories," for the Central Point Mixed Learning Strategy and the Random Example Selection Strategy.

A decision rule network is a graphical representation of decision rules generated at individual stages of a learning process, for a specific value of a given decision attribute. In our analysis, we considered two decision attributes, namely "minimum recommended height range" and "maximum recommended height range," each with seven values. Therefore, we prepared fourteen decision rule networks. A decision rule network shows decision rules and the relationships among them for different stages of the learning process. The nodes of the network represent the decision rules while the links in the network represent paths from one node to another, and the relationships between them. Using such a network, the numbers of rules and attributes can be easily determined for individual stages of the learning process. Fig. 2 shows a decision rule network for the same case as presented in Fig. 1.

A decision rule network can be used to study changes in attributes occurring during the learning process for a specific final decision rule produced at the last learning stage. This can be done using back-tracking, which starts with the final decision rule and follows links in the network. The links representing the smallest number of changes between adjacent stages are selected and placed in the path of the final decision rules. This enables us to determine the origin of each final decision rule and indicates the completion of the learning process for this rule. As an example, consider the path for the final decision rule: if $B=1, D=1, E=1, F=1$ then $K=1$, which implies that the recommended height range of a single-bay moment resisting frame is less than ten

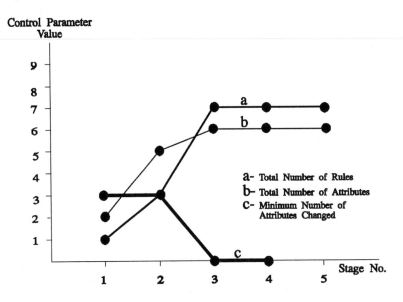

Figure 1. Learning Curves for Individual Control Parameters

stories. This decision rule was developed at the fifth stage of the learning process illustrated by the network shown in Fig. 2.

The analysis starts at the node representing the final decision, indicated in Fig. 2 by the shaded rectangle at stage five. All links for this node are considered, and the link with the smallest number of changes, in this case zero changes, is found for the fourth-stage node. This node is indicated by the shaded rectangle at the fourth stage. Again, all links for this node are analyzed and the link with the smallest number of changes, in this case zero changes, is found. The analysis continues through the network and the entire path is identified, as indicated by the solid line in Fig. 2. It shows that learning was completed in the third stage when the final decision rule was obtained. This rule has not been changed during the subsequent learning stages. This implies that not all 164 examples were necessary to generate such a rule. Our experience indicates that such a path is useful for improving our understanding of changes occuring in inductive learning and to obtain a better understanding of the final decision rules.

4. Knowledge Acquisition Process

Only the most significant results are discussed here. They were produced from 164 examples using five strategies of inductive learning and two strategies of example selection, as discussed in Section 3.

The knowledge acquisition process produced a number of correct decision rules which are generalizations of the examples used. These decision rules identify minimum

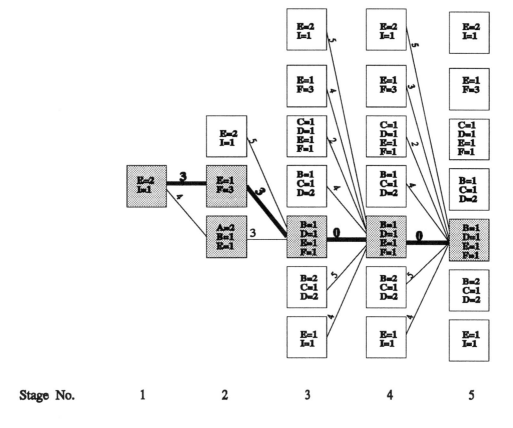

Stage No.   1   2   3   4   5

Figure 2.   Decision Rule Network for Minimum Recommended Height Range <10
            Stories (D=1)

height ranges for various groups of wind bracing types. For example, in the learning
process conducted using the central-point mixed learning strategy and the random
strategy of example selection, shown in Fig. 2, the following rule was obtained:

Rule No. 1: If B = 1, C = 1, D = 2 then K = 1 (less than ten stories)

In structural terms, this decision rule says that all wind bracings occupying the central
bay of the skeleton structure, and with a single horizontal truss on the top of bracing,
are feasible for heights less than ten stories.
    The principles of structural shaping of wind bracings are understood as directives
useful in the selection of individual bracing components for a given design case. These
principles are revealed by the changes that decision rule No. 1 undergoes when the
recommended height range is increased from "less than ten stories" to "ten to sixteen
stories." For the latter range, the inductive system produced three rules with the same
attributes as the first one. These rules and their structural interpretation are as follows:

Rule No. 2: If B = 1, C = 2, D = 2 then K = 2 (ten to sixteen stories)

All wind bracings occupying only the central bay of a skeleton structure with a single vertical truss and a single horizontal truss are feasible in the height range of ten to sixteen stories.

Rule No. 3: If B = 1, C = 1, D = 4 then K = 2 (ten to sixteen stories)

All wind bracings occupying only the central bay of a skeleton structure with three horizontal vertical trusses are feasible in the height range of ten to sixteen stories.

Rule No. 4: If B = 3, C = 1, D = 2 then K = 2 (ten to sixteen stories)

All wind bracings occupying three bays, with a single horizontal truss, are feasible in the height range of ten to sixteen stories.

Decision rule No. 1 was compared with these three decision rules (2 through 4) and the following structural shaping principles were discovered:

If you want to increase the recommended height range from less then ten stories to ten to sixteen stories for a one-bay bracing with a single horizontal truss, then use one of the following structural shaping principles:

1. Use a vertical truss in your bracing, or
2. Use three horizontal trusses, or
3. Use three-bay bracing.

Similar but more complex structural shaping principles were discovered for increasing the recommended height range from less than ten stories to sixteen to 24 stories. In this case, each principle is based on the change of two attributes. When the increase of recommended height range from less than ten stories to the 24-32-story range was analyzed and the decision rules compared, the most complex three-attribute structural shaping principles were discovered.

All of the decision rules and shaping principles discussed here are compiled in Fig. 3 in the form of two types of inductive learning decision trees. Fig. 3a provides a pictorial representation of the wind bracing types considered here. Fig. 3b gives individual decision rules. The decision tree has four levels corresponding to the recommended height ranges of less than ten, ten to sixteen, sixteen to 24 and 24 to 32 stories respectively. Decision rule No. 1, related to the recommended height range of less than ten stories, is placed on the bottom level. Decision rules 2 through 4, identifying wind bracings for the recommended height range of ten to sixteen stories, are on the level two. In the final tree the distance between adjacent levels is equal to one in terms of the number of attributes changed. In structural terms, when two adjacent levels are considered, a single wind bracing component must be changed or added to a type of wind bracing from the lower level, to move upward to the higher level. The developed inductive learning decision tree represents more than just the decision rules obtained. It also presents our improved understanding of the problem of structural shaping of wind bracings, which is in the form of the shaping principles discovered.

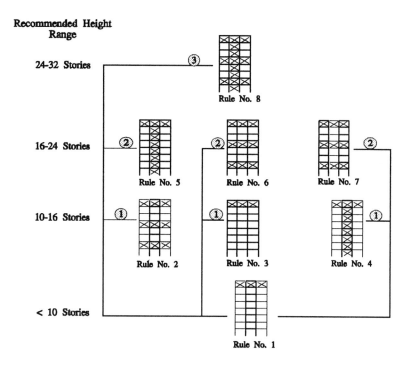

Figure 3a. Inductive Learning Decision Tree: Graphical Representation

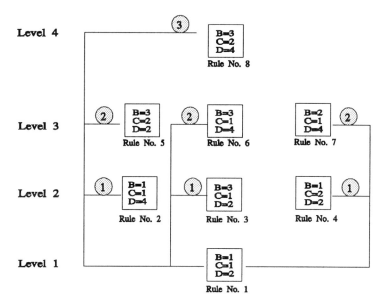

Figure 3b. Inductive Learning Decision Tree: Decision Rules

Unfortunately, not all of our decision rules were easy to interpret from the structural point of view. An example of such decision rule, shown in Fig. 2, is:

If $E = 2$ and $I = 1$ then $K = 1$.

This says: If there is a single horizontal truss system and there are no external joints, then the recommended height range is less than ten stories. The combination of attribute values $E = 2$ and $I = 1$ will never be used by a designer because it is incompatible, or illogical from the structural point of view. A wind bracing in the form of a belt truss system ($E = 2$) requires the presence of both external columns and external joints, and therefore the value of the I attribute must be other than 1. Initially, we considedred this rule false, and we could not explain why it had been generated. Only after a long analysis we concluded that it is a rare case of a rule which is logically true, but useless from the structural point of view. In the prepositional logic, decision rules are implications. In this particular instance, the implication has a true conclusion, even though the combination of its premises is false. Therefore, the logical value of the entire implication is "true." However, because of the incompatibility observed, this decision rule will never be executed, when it is included in a knowledge base. Usually, inductive learning produces the true structural rules, or implications, which have true both premises and conclusions.

The monitoring of the learning process revealed that it had not been completed. All learning control parameters were still undergoing changes when the learning was terminated due to an insufficient number of examples (Fig. 4). In Fig. 4, learning curves for different control learning parameters are given for two recommended heights and all learning strategies.

Fig. 4a shows learning curves for the total number of rules for the conclusion "recommended height range ten to sixteen stories". Both Figs. 4b and 4c present learning curves for the conclusion "recommended height range less than ten stories" Fig. 4b plots the total number of attributes, and Fig. 4c the minimum number of attributes changed at each stage of the learning process. The end of learning occurs when the learning curves become flat and their slopes approach zero. In our case, however, all learning curves were still showing changes, meaning that learning was in progress when terminated. Therefore, the results reported here are not final and could be different if more examples were available. Our examples were prepared by experts and simply reflected their subjective knowledge; this might be another reason why learning was relatively slow and more examples were needed.

5. Conclusions

The knowledge acquisition process produced interesting and useful results regarding the structural shaping of wind bracings in steel skeleton structures. Its practical impact is limited due to the incompleteness of the learning process. However, it has improved our understanding of the changes occurring in the inductive learning process, and has demonstrated its feasibility in structural engineering.

The results of the learning process were found to be only partially satisfactory. Not all the decision rules generated were correct from the structural point of view, and one of them included an infeasible combination of condition-attribute values. This disappointing result was in fact expected as the learning process was monitored. All

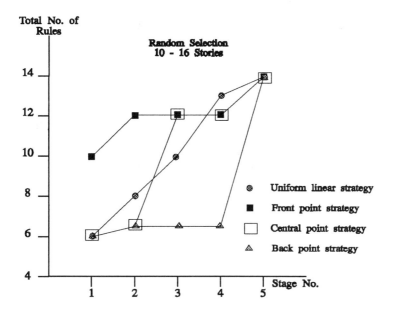

Figure 4a. Learning Curves for Control parameter: Total Number of Rules

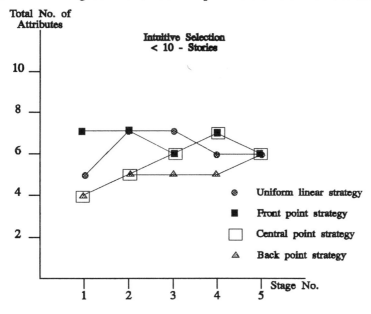

Figure 4b. Learning Curves for Control parameter: Total Number of Attributes

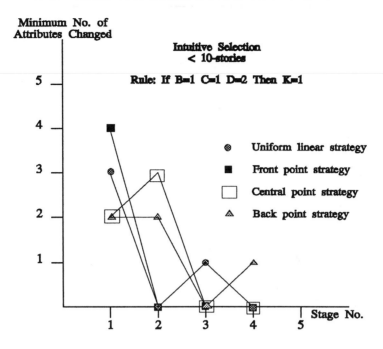

Figure 4c.   Learning Curves for Control parameter: Total Number of Attributes
              Changed

the learning control parameters indicated that the learning process had not been completed and that the results were not final, due to the insufficient number of examples used in our project. Decision rules containing infeasible combinations of condition attribute values would not have been produced, if domain-based constraints regarding these combinations had been imposed.

The entire project was reexamined to find a way to obtain more meaningful and sufficient decision rules, and to complete the knowledge acquisition process. Because no more examples of evaluated types of wind bracings could be obtained from the experts, we decided to generate more examples using the SODA computer software package. This activity is currently in progress. After its completion, the automated learning process and its evaluation will be attempted again.

We learned about the importance of examples and their adequacy in terms of their number and character. We also learned that a completely automated knowledge acquisition process in a complex civil engineering domain is still not possible. An inductive system is a potent knowledge acquisition tool, but its use still requires the involvement of a knowledge engineer with a strong domain background. This domain background is particularly important; it helps one to understand and interpret the often confusing learning results, it enables one to correct wrong results, and it is necessary to verify final results. Inductive learning enhances human learning, but does not eliminate it. A knowledge engineer is still central to automated knowledge acquisition, and will remain essential for the foreseeable future.

Acknowledgements

The authors are pleased to acknowledge the cooperation of Voytech Systems, Inc. of Regina, Canada, and of Waterloo Engineering Software Inc. of Waterloo, Canada. Voytech Systems, Inc. donated the inductive system ROUGH used in our research, and Waterloo Engineering Software Inc. provided an excellent technical support in the preparation of the SODA-generated design examples.

Appendex-References

Arciszewski T., (1985). "Decision Making Parameters and their Computer-Aided Analysis for Wind Bracings in Steel Skeleton Structures," *Advances in Tall Buildings*, Van Nostrand Publishing Company.

Arciszewski T., Mustafa M., (1989). "Inductive Learning Process: The User's Perspective," in *Machine Learning*, edited by R. Forsyth, Chapman and Hall.

Arciszewski T., Mustafa M., Ziarko W., (1987). "A Methodology of Design Knowledge Acquisition for Use in Learning Expert Systems," *International Journal of Man-Machine Studies*, No. 27.

Arciszewski T., Olowokere D., Alkhatib M., (1988). "Modeling of Belt-Truss Bracing Systems in Steel Skeleton Structures," *Proceedings of the International Tall Buildings Conference*, Shanghai, China, April.

Arciszewski, T., Ziarko, W., (1990). "Inductive Learning in Civil Engineering: Rough Sets Approach," *Microcomputers in Civil Engineering*, Vol. 26, No. 3.

Arciszewski T., Ziarko W., (1987). "Adaptive Expert System for Preliminary Engineering Design," *Revue Internationale D.E. CFAO et D'Intographie*, Vol. 2, No. 1.

Arciszewski T., Ziarko W., (1988). "Adaptive Expert System for Preliminary Design of Wind Bracings in Skeleton Structures," *Second Century of Skyscrapers*, van Nostrand Publishing Company.

Maher, M., (1984) "HI-RISE : A Knowledge-Based Expert System for Preliminary Design of High Rise Buildings," Ph.D. Thesis, Carnegie-Mellon University, Pittsburgh, PA.

Mustafa, M., Arciszewski, T., (1989). "Knowledge Acquisition: Engineering Methodology of Inductive Learning", *Proceedings of Workshop on Knowledge Acquisition, Joint International Conference on Artificial Intelligence*, Detroit.

Scalzi, J. B., (1981). "Drift in High-Rise Buildings," In Lin, T. Y., Stotesbury, S. D., *Structural Concepts and Systems for Architects and Engineers*, John Wiley & Sons, Inc.

Schueller, W., (1976). *High-Rise Bulding Structures*, John Wiley & Sons, New York.

Sriram, D., (1987). Knowledge Based Applications for Structural Design, *Computational Mechanics Publications*.

Taranath, B. S., (1988). *Structural Analysis and Design of Tall Buildings*, McGraw-Hill Book Company, New York.

CHAPTER 11

Text and Reference Books on Knowledge Acquisition and Machine Learning

Yoram Reich

1. Introduction

Knowledge acquisition and machine learning are emerging technologies for the transfer of knowledge from various sources into a form that can be readily used to solve problems. Over the past several years there has been a growing interest in studying the processes involved in these technologies, leading to the proliferation of publications. Text books that cover these disciplines are still rare and there is none that is specifically directed at engineering applications.

The lack of substantial work in applying advanced knowledge acquisition and machine learning techniques to engineering problems is apparent in the references cited in this review. Nevertheless, this chapter is intended to be an aid for instructors, researchers, and practitioners in selecting reading material, from an introductory to an advanced level. It is hoped that better acquaintance with the literature will inspire researchers to initiate new research projects and encourage engineers to use these techniques in their work. Furthermore, these references should be useful not only to those working on computer-support systems but also to others: teachers, students, and other engineers; to some extent the engineering community as a whole is involved in knowledge transfer processes of some form.

This review is subdivided into two sections. The first section contains a description of books that can serve as a main source of information on knowledge acquisition and machine learning. The second section contains a list of additional references that are more focussed in specialized domains.

2. Texts and References

This section includes a brief description of text and reference books. The description of each book contains general comments on the presentation of the material, the level of the book (introductory, intermediate, advanced), the intended audience (e.g., researchers, engineers, general, etc.). In most cases, the content of the book or its summary appears at the end of the description. First, manual knowledge acquisition techniques are described, followed by computer-supported knowledge acquisition techniques. Finally, automatic techniques (i.e., machine learning) are detailed.

Knowledge Acquisition Principles and Guidelines, K. L. McGraw & K. Harbison-Briggs, Prentice Hall, 1989.

Comments: This book provides an excellent and detailed account of knowledge acquisition, allowing the appreciation of the complexity of knowledge acquisition. The book deals with this complexity systematically by following the methodology of systems engineering. The book identifies the major components of the knowledge acquisition process and addresses them in separate chapters as seen in the content bellow. Many techniques for carrying out each step in the knowledge acquisition process are described, and their benefits and weaknesses are articulated.

Level: Introductory to advanced

Intended audience: General

Contents:

- Knowledge acquisition for expert systems
- Systems engineering perspective on the knowledge acquisition process
- Organizing a knowledge acquisition program
- Working effectively with domain experts
- Conceptualizing the domain
- Structuring the knowledge acquisition process
- Interviewing for content and clarification
- Tracing the decision making process to acquire
- Acquiring knowledge from multiple experts
- Aids and tools for knowledge acquisition
- Evaluating knowledge base development efforts

Knowledge Acquisition for Expert Systems, A. Hart, McGraw-Hill, 1986.

Comments: This book provides a simplified account of knowledge acquisition as opposed to the comprehensive view presented in the previous book. It is less comprehensive as the previous book. The level at which material is presented may help undergraduate students who wish to get a brief exposure to the area.

Level: Introductory

Intended audience: Students

Contents:

- The nature of expertise
- Programs as experts
- Systems analysis – a comparison
- The knowledge engineer
- Fact-finding by interviews
- Reasoning and probability theory
- Fuzziness in reasoning
- Machine induction
- The repertory grid

Knowledge Acquisition for Knowledge-Based Systems, Knowledge-Based Systems Book Series, Vol. 1, B. R. Gaines and J. Boose, Academic Press, 1988.

Knowledge Acquisition Tools for Expert Systems, Knowledge-Based Systems Book Series, Vol. 2, J. Boose and B. R. Gaines, Academic Press, 1988.

Machine Learning and Uncertain Reasoning, Knowledge-Based Systems Book Series, Vol. 3, B. R. Gaines and J. Boose, Academic Press, 1990.

The foundations of Knowledge Acquisition, Knowledge-Based Systems Book Series, Vol. 4, J. Boose and B. R. Gaines, Academic Press, 1990.

Comments: Each book in this series contains a collection of previously published papers. Most of the papers appeared in *The International Journal of Man-Machine Studies*. Beside the structuring of the studies into four volumes, there has not been an attempt to further structure them into more detailed topics. Nevertheless, these books serve as a good source of references on the state-of-the-art of knowledge acquisition techniques and tools.

Level: Introductory to advanced

Intended audience: General

Expertise Transfer for Expert System Design, J. H. Boose, Elsevier, 1986.

Comments: This book describes a knowledge acquisition system called ETS. It provides a background for a large body of knowledge acquisition research based on Personal Construct Psychology of Kelly (1955). The book highlights many issues in knowledge acquisition, all in the context of Kelly's theory. Furthermore, the book details many examples of the possible applications this theory can support.

Level: Introductory to intermediate

Intended audience: General

Contents:

- Introduction and background
- Expertise transfer system description
- Applications of ETS
- Appendices (includes several detailed applications)

Automatic Knowledge Acquisition for Expert Systems, S. Marcus (Ed.), Kluwer Academic Publishers, 1988.

Comments: This book contains a collection of research reports on knowledge acquisition tools. Each tool is devoted to a specific problem-solving paradigm. This thrust of research is based on McDermott's view of task-specific knowledge acquisition tools. Such effort is a first step at understanding the kinds of knowledge acquisition tools and machine learning techniques that can assist in acquiring knowledge for the variety of tasks in engineering.

Level: Intermediate to advanced

Intended audience: Researchers

Contents:

- Introduction
- MORE: From observing knowledge engineers to automatic knowledge acquisition

- MOLE: A knowledge-acquisition tool for cover-and-differentiating systems
- SALT: A knowledge-acquisition tool for propose-and-revise systems
- KNACK: Sample-driven knowledge acquisition for reporting systems
- SIZZLE: A knowledge-acquisition tool specialized for the sizing task
- RIME: preliminary work toward a knowledge-acquisition tool
- Preliminary steps toward a taxonomy of problem-solving methods

Knowledge Acquisition: Selected Research and Commentary, S. Marcus (Ed.), Kluwer Academic Publishing, 1990.

Comments: This book has appeared as a special issue of *Machine Learning, Vol 4 (3/4)*. The book contains three original papers and a list of commentaries, written by leading researchers, on knowledge acquisition and its automation by computer tools.

Level: Advanced

Intended audience: Researchers and Engineers

Introduction to Machine Learning, Y. Kodratoff, Morgan Kaufmann, 1988.

Comments: This book is the only graduate level textbook on machine learning available. It is a comprehensive treatment of several of the main topics in machine learning: similarity-based learning, explanation-based learning, and analogy. The discussion is formal and grounded in theory. Examples and exercises are included on each of the topics.

Level: Advanced

Intended audience: Researchers and Engineers

Contents:

- Why machine learning and AI
- Theoretical foundations for machine learning
- Representation of complex knowledge by clauses
- Representation of knowledge about actions and the addition of new rules to a knowledge base
- Learning by doing
- A formal representation of Version Spaces
- Explanation-based learning
- Learning by similarity detection: the empirical approach
- Learning by similarity detection: the 'rational' approach
- Automatic construction of taxonomies: techniques for clustering
- Debugging and understanding in depth: the learning of micro-worlds
- Learning by analogy

Genetic Algorithms in Search, Optimization and Machine Learning, D. E. Goldberg, Addison-Wesley, 1989.

Comments: This is the first textbook on genetic algorithms. The book starts with a gentle introduction leading to the state-of-the-art of the field. Pascal computer code is provided in the appendix and exercises are given after each chapter. The exercises test the application of the code, as well its potential improvements. The book points to many applications of genetic algorithms but does not study any application in detail.

Level: Introductory to advanced

Intended audience: General

Contents:

- A gentle introduction to genetic algorithms
- Genetic algorithms revisited: Mathematical foundations
- Computer implementation of a genetic algorithm
- Some applications of genetic algorithms
- Advanced operators and techniques in genetic search
- Introduction to genetic-based machine learning
- Applications of genetic-based machine learning

Computer Systems That Learn, S. M. Weiss and C. A. Kulikowski, Morgan Kaufmann, 1990.

Comments: This book provides an excellent introduction to the similarity-based learning paradigm. The book describes methods for evaluation learning techniques. The review of testing techniques is comprehensive and valuable to anyone who wishes to evaluate such techniques. The discussion of learning techniques concentrates on three learning paradigms: statistical pattern recognition, neural networks, and symbolic machine learning. The description of the three paradigms includes background, and applied material, as well as current research issues. A discussion on the relative advantages of the three paradigms follows their description. The book concludes with a chapter discussing the interaction between machine learning and expert systems.

Level: Introductory to advanced

Intended audience: General

Contents:

- Overview of machine learning
- How to estimate the true performance of a learning system
- Statistical pattern recognition
- Neural nets
- Machine learning: easily understood decision rules
- Which technique is best
- Expert systems

Machine Learning: An Artificial Intelligence Approach, R. S. Michalski, J. G. Carbonell, & T. M. Mitchell (Eds.), Tioga Publishing Company, 1983.

Comments: This is the first collection of research studies in machine learning. The papers are divided into major paradigms of machine learning as existed in 1983. A preface containing an introduction to the papers highlights their contribution. The book contains extensive bibliography on machine learning published until 1983. The bibliography is classified along several dimensions to facilitate effective indexing.

Level: Advanced

Intended audience: Researchers and engineers

Contents:

- General issues in machine learning
- Learning from examples
- Learning in problem-solving and planning
- Learning from observation and discovery
- Learning from instruction
- Applied leaning systems

Machine Learning: An Artificial Intelligence Approach, Vol 2, R. S. Michalski, J. G. Carbonell, & T. M. Mitchell (Eds.), Morgan Kaufmann, 1986.

Comments: This is the second book in the series. It contains a collection of papers classified into paradigms that differ from the first book. The first part contains an introduction on the various techniques; it highlights the potential of machine learning. The style of presentation is similar to the first book.

Level: Advanced

Intended audience: Researchers and engineers

Contents:

- General issues
- Learning concepts and rules from examples
- Cognitive aspects of learning
- Learning by analogy
- Learning by observation and discovery
- An exploration of general aspects of learning

Machine Learning: An Artificial Intelligence Approach, Vol 3, Y. Kodratoff & R. S. Michalski (Eds.), Morgan Kaufmann, 1990.

Comments: This is the third book of the series. The classification of the papers is based on the current view of the field and is reflected in the content. In particular, the content points to a new focus on integrated and heterogeneous learning systems. Such systems have a potential to support real engineering applications. The comprehensive bibliography augments those printed in the first two volumes.

Level: Advanced

Intended audience: Researchers and engineers

Contents:

- General issues
- Empirical learning methods
- Analytical learning methods
- Integrated learning systems
- Subsymbolic and heterogeneous learning systems
- Formal analysis

Machine and Human Learning: Advances in European Research, Y. Kodratoff & A. Hutchinson (Eds.), GP Publishing, 1989.

Comments: This book contains papers on machine learning classified along several dimensions: rule learning programs, the use of advanced representations for learning, the use of numerical techniques such as statistics, and learning through human-computer interaction. The book is similar in nature to the three previous volumes, except for the lack of extensive bibliography collection.

Level: Advanced

Intended audience: Researchers and engineers

KARDIO: A Study in Deep and Qualitative Knowledge for Expert Systems, I. Bratko, I. Mozetič & N. Lavrač, The MIT Press, 1989.

Comments: This book demonstrates how knowledge can be transferred between two representations to partially overcome the bottleneck of knowledge acquisition. Knowledge initially coded as a causal model, is used to generate cases that are used by a machine learning program to generate rules. The inefficient, but accurate knowledge, transfers to efficient, but heuristic knowledge. This approach can be very appealing in engineering where models of artifacts or processes are relatively easy to construct.

Level: Intermediate to Advanced

Intended audience: Researchers and engineers

Contents:

- Introduction
- Qualitative model of the heart
- Model interpretation and derivation of surface knowledge
- Knowledge base compression by means of machine learning
- Further developments: hierarchies and learning of models

Readings in Machine Learning, J. Shavlik & T. Dietterich (Eds.), Morgan Kaufmann, 1990.

Comments: This is a very comprehensive collection of machine learning references. It spans all areas of research and provides extensive introductory material. The book is good as a source for graduate course on machine learning and as reference material for researchers and engineers.

Level: Introductory to advanced

Intended audience: General

Contents:

- General aspects of machine learning
- Inductive learning using pre-classified training examples
- Unsupervised concept learning and discovery
- Improving the efficiency of a problem solver
- Using pre-existing domain knowledge inductively
- Explanatory/inductive hybrids

Parallel Distributed Processing: Explorations in the Microstructure of Cognition, Vol 1: Foundation and Vol 2: Psychological & Biological Models, D. E. Rumelhart, J. L. McClelland, & the PDP Research Group, The MIT Press, 1986.

Comments: These are the classical books on neural networks. They provide an in-depth summary of the origins of neural networks, and their potential benefits. They contains a description of the various mechanisms and their formal analysis. The second book, which describe the application of neural networks to explain psychological or biological phenomena can serve as a source of case studies on designing networks to satisfy certain objectives.

Level: Intermediate

Intended audience: General

Contents:
- The PDP perspective
- Basic Mechanisms
- Psychological processes
- Biological mechanisms

Machine Learning Applications in Expert Systems and Information Retrieval, R. Forsyth & R. Rada, Ellis Horwood, 1986.

Comments: This book contains two separate parts. The first part introduces machine learning through a brief review of neural networks and other numeric learning techniques, symbolic learning and genetic algorithms. This introduction is detailed but basic. The second part describes the role of learning in information retrieval systems. First, this part reviews existing studies and then it provides a detail case study of a specific system. This part should be very important to engineers since technical information systems are part of any engineering project.

Level: Introductory to intermediate

Intended audience: General

Contents:
- Introduction to machine learning
- Black-box methods
- Learning structural descriptions
- Evolutionary learning strategies
- Towards the learning machine
- A theme for machine learning in information retrieval
- knowledge-sparse learning
- Knowledge-rich learning
- The MEDLARS system

Machine Learning, Principles and Techniques, Forsyth, R. (Ed.), Chapman and Hall, 1989.

Comments: This book contains a diverse collection of 13 chapters divided into four parts as noted below. The chapters span issues in symbolic learning, genetic algorithms and neural networks and can serve as a gentle introduction to these topics.

Level: Basic to Intermediate

Intended audience: General

Contents:
- Background

- Biologically inspired systems
- Automated Discovery
- Long-Term Perspectives

Classification and Regression Trees, L. Breiman, J. H. Friedman, R. A. Olshen, & C. J. Stone, Wadsworth, 1984.

Comments: This book provides an in-depth description of tree structured rules for classification. The book provides a sound theory for the algorithms developed and illustrates them in two domains. The scientific methodology of experimenting with the algorithms is useful to any researcher engaged in the development or use of machine learning.

Level: Advanced

Intended audience: Researchers

Contents:

- Tree structured methodology in classification
- Examples of trees used in classification
- Use of trees in regression
- Theoretical framework for tree structures methods

3.   Bibliography list of additional texts and references

This section includes an incomplete list of additional text and reference books. Comprehensive lists of article, conference papers, and other publications can be found in (Michalski et al. 1983; Michalski et al. 1986; Kodratoff and Michalski, 1990). Among the additional sources are the proceedings of the (1) International Conference/Workshop on Machine Learning, (2) Knowledge Acquisition for Knowledge-Based Systems Workshop, (3) International Conference on Genetic Algorithms, (4) National Conference on Artificial Intelligence, (5) International Joint Conference on Artificial Intelligence, (6) International Symposium on Methodologies of Intelligent Systems, (7) European Conference on Artificial Intelligence, and (8) European Working Session on Learning. Other references can be found in the following incomplete list of journals: (1) Machine Learning, (2) Knowledge Acquisition, (3) International Journal of Man-Machine Studies, (4) Artificial Intelligence, (5) Neural Network Journal, (6) IEEE Transactions on Systems, Man and Cybernetics, and (7) IEEE Transactions on Pattern Analysis and Machine Intelligence.

Bareiss, R. (1989). *Exemplar-Based Knowledge Acquisition*, Academic Press, Boston, MA.

Bolc, L., editor (1987). *Computational Models of Learning*, Springer-Verlag, Berlin.

Brule, J. and Blount, A. (1989). *Knowledge Acquisition*, McGraw-Hill, New York.

Diaper, D. (1989). *Knowledge Elicitation: Principles, Techniques, and Applications*, John Wiley & Sons, Somerset, N.J.

Fisher, D. and Pazzani, M., editors (1991). *Computational Approaches to Concept Formation*, Morgan Kaufmann, San Mateo, CA.

Ginsberg, A. (1988). *Automatic Refinement of Expert System Knowledge Bases*, Morgan Kaufmann, San Mateo, CA.

Glymour, C., Scheiner, R., Spirtes, P., and Kelly, K. (1987). *Discovering Causal Structure: Artificial Intelligence, Philosophy of Science, and Statistical Modeling*, Academon Press, Orlando. Contains: forward by Simon, TETDAD project.

Gruber, T. R. (1989). *The Acquisition of Strategic Knowledge*, Academic Press, Boston, MA.

Hammond, K. J. (1989). *Cased-Based Planning*, Academic Press, Boston, MA.

Kidd, A., editor (1987). *Knowledge Elicitation for Expert Systems: A Practical Handbook*, Plenum Press, New York.

Laird, J. E., Rosenbloom, P. S., and Newell, A. (1986). *Universal Subgoaling and Chunking: The Automatic Generation and Learning of Goal Hierarchies*, Kluwer, Hingman, MA.

Langley, P., Simon, H. A., Bradshaw, G. L., and Zytkow, J. M. (1987). *Scientific Discovery: Computational Explorations of the Creative Processes*, MIT Press, Cambridge, MA.

McClelland, J. L. and Rumelhart, D. E. (1988). *Explorations in Parallel Distributed Processing: A Handbook of Models, Programs, and Exercises*, The MIT Press, Cambridge, MA. (Includes a diskettes with programs).

McDermott, J. (1988). "Automating the knowledge acquisition process.". Video Lecture, Morgan Kaufmann, San Mateo, CA.

Meyer, M. and Booker, J. (1991). *Eliciting and Analyzing Expert Judgement: A Practical Guide*, Knowledge-Based Systems Book Series, Vol. 5, Academic Press, Boston, MA.

Mooney, R. J. (1990). *A General Explanation-Based Learning Mechanism and its Application to Narrative Understanding*, Morgan Kaufmann, San Mateo, CA.

Morik, K., editor (1989). *Knowledge Representation and Organization in Machine Learning*, Lecture Notes in Artificial Intelligence, Vol. 347, Springer-Verlag, Berlin.

Muggleton, S. (1990). *Inductive Acquisition of Expert Knowledge*, Addison-Wesley, Reading, MA.

Musen, M. A. (1989). *Automated Generation of Model-Based Knowledge-Acquisition Tools*, Morgan Kaufmann, San Mateo, CA.

Pazzani, M. J. (1990). *Creating a Memory of Causal Relationships: An Integration of Empirical and Explanation-Based Learning Methods*, Lawrence Erlbaum, Hillsdale, N.J. (Includes a diskette with programs).

Rumelhart, D. (1990). "Brainstyle computation and learning.". Video Lecture, Morgan Kaufmann, San Mateo, CA.

Scott, A. C., Clayton, J. E., and Gibson, E. L. (1991). *A Practical Guide to Knowledge Acquisition*, Addison-Wesley, Reading, MA.

Shapiro, A. D. (1987). *Structured Induction in Expert Systems*, Addison-Wesley, Wokingham, England.

Shavlik, J. W. (1990). *Extending Explanation-Based Learning by Generalizing the Structure of Explanations*, Morgan Kaufmann, San Mateo, CA.

Shaw, M. L. G., editor (1981). *Recent Advances in Personal Construct Technology*, Academic Press, London.

Shortliffe, E. (1988). "Knowledge acquisition: Building and maintaining the knowledge base for expert systems.". Video Lecture, Morgan Kaufmann, San Mateo, CA.

Shrager, J. and Langley, P., editors (1990). *Computational models of scientific discovery and theory formation*, Morgan Kaufmann, San Mateo, CA.

Acknowledgments
This review uses the format of (Garrett 1989; "Text and reference books on expert systems and artificial intelligence." In Mohan, S. and Maher, M. L., editors, *Expert Systems for Civil Engineers: Education*, pages 82–94, American Society of Civil Engineers, New York, NY.). I would like to thank the following publishers who provided material and permission to reprint the table of content of their books: Academic Press, Addison-Wesley, Elsevier, Kluwer Academic Publishers, McGraw-Hill, MIT Press, Morgan Kaufmann, North-Holland, Prentice Hall. This work has been supported in part by the Engineering Design Research Center, a National Science Foundation Engineering Research Center, and the Sun Company Grant for Engineering Design Research.

SUBJECT INDEX
Page number refers to first page of paper.

AUTHOR INDEX
Page number refers to first page of paper.